Essential Oils

Recent Titles in
Q&A Health Guides

Depression: Your Questions Answered
Romeo Vitelli

Food Labels: Your Questions Answered
Barbara A. Brehm

Smoking: Your Questions Answered
Stacy Mintzer Herlihy

Teen Stress: Your Questions Answered
Nicole Neda Zamanzadeh and Tamara D. Afifi

Grief and Loss: Your Questions Answered
Louis Kuykendall Jr.

Healthy Friendships: Your Questions Answered
Lauren Holleb

Trauma and Resilience: Your Questions Answered
Keith A. Young

Vegetarian and Vegan Diets: Your Questions Answered
Alice C. Richer

Yoga: Your Questions Answered
Anjali A. Sarkar

Teen Pregnancy: Your Questions Answered
Paul Quinn

Sexual Harassment: Your Questions Answered
Justine J. Reel

Sports Injuries: Your Questions Answered
James H. Johnson

ESSENTIAL OILS

Your Questions Answered

Randi Minetor

Q&A Health Guides

GREENWOOD

An Imprint of ABC-CLIO, LLC
Santa Barbara, California • Denver, Colorado

Library of Congress Cataloging-in-Publication Data
Names: Minetor, Randi, author.
Title: Essential oils : your questions answered / Randi Minetor.
Description: Santa Barbara, California : Greenwood, [2022] | Series: Q&A
    health guides | Includes bibliographical references and index. |
    Summary: "This book helps separate myths from facts and discusses the
    role essential oils play in people's personal, professional, and
    spiritual lives"— Provided by publisher.
Identifiers: LCCN 2021035821 (print) | LCCN 2021035822 (ebook) |
    ISBN 9781440877841 (hardcover) | ISBN 9781440877858 (ebook)
Subjects: LCSH: Essences and essential oils. | Essences and essential
    oils—Miscellanea.
Classification: LCC QD416 .M63 2022 (print) | LCC QD416 (ebook) |
    DDC 615.3/21—dc23
LC record available at https://lccn.loc.gov/2021035821
LC ebook record available at https://lccn.loc.gov/2021035822

ISBN: 978-1-4408-7784-1 (print)
      978-1-4408-7785-8 (ebook)

26 25 24 23 22   1 2 3 4 5

This book is also available as an eBook.

Greenwood
An Imprint of ABC-CLIO, LLC

ABC-CLIO, LLC
147 Castilian Drive
Santa Barbara, California 93117
www.abc-clio.com

This book is printed on acid-free paper ∞

Manufactured in the United States of America

Contents

Series Foreword

All of us have questions about our health. Is this normal? Should I be doing something differently? Whom should I talk to about my concerns? And our modern world is full of answers. Thanks to the internet, there's a wealth of information at our fingertips, from forums where people can share their personal experiences to Wikipedia articles to the full text of medical studies. But finding the right information can be an intimidating and difficult task—some sources are written at too high a level, others have been oversimplified, while still others are heavily biased or simply inaccurate.

Q&A Health Guides address the needs of readers who want accurate, concise answers to their health questions, authored by reputable and objective experts, and written in clear and easy-to-understand language. This series focuses on the topics that matter most to young adult readers, including various aspects of physical and emotional well-being as well as other components of a healthy lifestyle. These guides will also serve as a valuable tool for parents, school counselors, and others who may need to answer teens' health questions.

All books in the series follow the same format to make finding information quick and easy. Each volume begins with an essay on health literacy and why it is so important when it comes to gathering and evaluating health information. Next, the top five myths and misconceptions that surround the topic are dispelled. The heart of each guide is a collection of questions and answers, organized thematically. A selection of five case studies provides real-world examples to illuminate key concepts. Rounding out each volume are a directory of resources, glossary, and index.

It is our hope that the books in this series will not only provide valuable information but will also help guide readers toward a lifetime of healthy decision making.

Acknowledgments

Many thanks to the hundreds of sources who contributed their expertise to this book through the peer-reviewed academic and professional journals and books that chronicle their work.

To my editor, Maxine Taylor, and the production staff at ABC-CLIO, I thank you for the opportunity to explore this topic and for the excellent job you always do on your books.

I am ever grateful for the support and encouragement of my dear friends Ken Horowitz and Rose-Anne Moore, Lisa Jaccoma, Kevin Hyde, Martha and Peter Schermerhorn, Martin A. Winer, Ruth Watson and John King, Cindy Blair, cousins Paula and Rich Landis, and the remarkable number of people who seek me out online to comment favorably on my work. This is my 63rd book written under my own name and my 80th overall, and it is tremendously gratifying to know that a few of them have had a positive impact.

And to Nic Minetor, my husband, with whom I shared the 14-month pandemic in quiet (and healthy) harmony, thank you for the meals you cooked, the repairs you made, the breaks we took together, the shelves you installed, and all the other things you do while I'm writing that I often don't notice or appreciate until much later. The next book, I promise, will take us back out on the road.

Introduction

Before penicillin transformed the way doctors treated infections, before aspirin became the readily accessible drug to counteract all kinds of pain, and before X-rays provided insights about injuries under the skin, the world of science-based medicine competed with salespeople on street corners, hawking concoctions that we now deride as "snake oil."

These charlatans emerged in early eighteenth-century England and soon spread throughout Europe, selling various forms of liquid as a cure-all for everything from pox to prostate issues. As towns began to grow on the American continents, snake oil salespeople crossed the Atlantic Ocean to take advantage of the people there, and soon unscrupulous colonists saw the moneymaking opportunity and duplicated the scam. A few swigs of whatever was in the bottle—which could vary from diluted mineral oil to petroleum-based mixtures, water, urine, and occasionally a bit of oil derived from actual rattlesnakes—was touted to cure hair loss, agues, agita, flu, muscle aches, hearing loss, blindness, and any other mysterious malady a customer could describe.

Despite their ability to secure patents—generating the nickname "patent medicine" as a catch-all for anything in a bottle—these substances remained outside of the bounds of science for centuries. Laboratory testing was out of the question, as most communities had no scientific labs; even if they had, salespeople would not have been eager to turn over bottles of their fraudulent elixirs for examination by professionals. With no legitimate cures for most ailments available in remote communities or rural areas, snake oils became the only option, so salespeople took people's money and handed them containers of foul-smelling and worse-tasting "medicine" that did nothing for them at all.

Pharmaceuticals were not regulated in the United States until the 1906 Food and Drugs Act. This law required food and medicine labeling to include a list of the actual ingredients, as well as formal testing of any substance about which the manufacturer made any kind of health claim. With the act in place, fake medicines faded from the marketplace as scientists proved that they had none of the positive effects they touted. Some simply vanished of their own accord, because their originators knew that their swill could not possibly hold up against legitimate research. It seemed that the era of miracle cure-alls had come to an end, though the term "snake oil" stuck, a handy moniker for any substance that claimed to have medicinal value, but that could not prove that it did so much as quench thirst.

Into this environment of age-old distrust and skepticism came a new set of miracles: essential oils. Touting their natural origins, purity, and pleasant scents, these liquids have gained tremendous popularity since the birth of aromatherapy in 1937, most recently through multilevel marketing companies that allow consumers to handle the oils, inhale their scents, and learn how to apply them properly before they buy them. Today essential oils are found in supermarkets, drug stores, bed-and-bath stores, online apothecaries, and even directly from farms that grow the herbs, spices, and other plants used to acquire them.

Many purveyors of essential oils make what seem like outlandish claims about their antimicrobial properties, telling consumers that these natural oils can substitute for the drugs manufactured by pharmaceutical companies. Some recommend dosing oneself with the oils by placing a drop or two in a glass of water daily, while some even suggest sprinkling them on food. Every essential oil marketer touts their effectiveness in altering mood through aromatherapy, using an electric device to vaporize the oils so that the user can inhale their essence. Some consumers swear by the results. Until recently, however, little to no scientific research existed to back up any claim these marketing companies make. For decades, essential oils might just as well have been the snake oils of old, with nothing but anecdotal evidence to suggest that they had any effect, positive or negative, on the human body.

In the past few years, however, essential oils have taken some first tentative steps out of the shadows cast by patent medicine predecessors and into legitimacy. Today the food manufacturing and packaging industry sponsors many studies to determine if essential oils have any of the antimicrobial properties their marketers claim, in hopes of using them as natural preservatives for organic foods. The early results have been promising, as you will see detailed in this book. Some essential oils are already in use for this purpose, while the industry has embraced others as natural flavors in foods, especially packaged baked goods.

If some essential oils can combat mold, bacteria, and fungus in food, it seems likely that they could have a role in doing the same in other settings: in cleaning the home, for example, and even in the human body. Research continues to try to determine if this could be the case; you will find information about some of these studies in the pages that follow.

Whether or not essential oils turn out to be the natural wonders that consumers seek, users apply drops of essential oils to diffusers to scent the air in their homes and offices as a way to combat stress and boost their mood and as an aid to calming meditation and prayer. Many people also use them as if they were medicine, mixing them with carrier oils and applying them topically or taking them internally. This book does not provide recipes for blending essential oils for various purposes, but it does guide readers to check safety guidelines for the oils' use and to pay attention to science rather than salespeople about how best to employ the oils in daily life.

This is my third book about essential oils and the role they may play in your and your family's personal, professional, and spiritual lives. If you choose to explore further, I recommend my two earlier books: *Essential Oils and Aromatherapy: An Introductory Guide* (Sonoma Press, 2014) and *Essential Oils of the Bible: Connecting God's Word to Natural Healing* (Althea Press, 2016).

With some essential oils making the leap from snake oil to science, there may be discoveries on the horizon that will have an impact on the massive and growing market for them. This book will help you separate myths from facts while pointing you toward science and researchers who lead the field in determining whether these oils do indeed have a positive impact in fighting disease or infection, or whether their sole benefit is their pleasant scent.

Guide to Health Literacy

On her 13th birthday, Samantha was diagnosed with type 2 diabetes. She consulted her mom and her aunt, both of whom also have type 2 diabetes, and decided to go with their strategy of managing diabetes by taking insulin. As a result of participating in an after-school program at her middle school that focused on health literacy, she learned that she can help manage the level of glucose in her bloodstream by counting her carbohydrate intake, following a diabetic diet, and exercising regularly. But, what exactly should she do? How does she keep track of her carbohydrate intake? What is a diabetic diet? How long should she exercise and what type of exercise should she do? Samantha is a visual learner, so she turned to her favorite source of media, YouTube, to answer these questions. She found videos from individuals around the world sharing their experiences and tips, doctors (or at least people who have "Dr." in their YouTube channel names), government agencies such as the National Institutes of Health, and even video clips from cat lovers who have cats with diabetes. With guidance from the librarian and the health and science teachers at her school, she assessed the credibility of the information in these videos and even compared their suggestions to some of the print resources that she was able to find at her school library. Now, she knows exactly how to count her carbohydrate level, how to prepare and follow a diabetic diet, and how much (and what) exercise is needed daily. She intends to share her findings with her mom and her aunt, and now she wants to create a chart that summarizes what she has learned that she can share with her doctor.

Samantha's experience is not unique. She represents a shift in our society; an individual no longer views himself or herself as a passive recipient of medical care but as an active mediator of his or her own health. However, in this era when any individual can post his or her opinions and experiences with a particular health condition online with just a few clicks or publish a memoir, it is vital that people know how to assess the credibility of health information. Gone are the days when "publishing" health information required intense vetting. The health information landscape is highly saturated, and people have innumerable sources where they can find information about practically any health topic. The sources (whether print, online, or a person) that an individual consults for health information are crucial because the accuracy and trustworthiness of the information can potentially affect his or her overall health. The ability to find, select, assess, and use health information constitutes a type of literacy—health literacy—that everyone must possess.

THE DEFINITION AND PHASES OF HEALTH LITERACY

One of the most popular definitions for health literacy comes from Ratzan and Parker (2000), who describe health literacy as "the degree to which individuals have the capacity to obtain, process, and understand basic health information and services needed to make appropriate health decisions." Recent research has extrapolated health literacy into health literacy bits, further shedding light on the multiple phases and literacy practices that are embedded within the multifaceted concept of health literacy. Although this research has focused primarily on online health information seeking, these health literacy bits are needed to successfully navigate both print and online sources. There are six phases of health information seeking: (1) Information Need Identification and Question Formulation, (2) Information Search, (3) Information Comprehension, (4) Information Assessment, (5) Information Management, and (6) Information Use.

The first phase is the *information need identification and question formulation phase*. In this phase, one needs to be able to develop and refine a range of questions to frame one's search and understand relevant health terms. In the second phase, *information search*, one has to possess appropriate searching skills, such as using proper keywords and correct spelling in search terms, especially when using search engines and databases.

It is also crucial to understand how search engines work (i.e., how search results are derived, what the order of the search results means, how to use the snippets that are provided in the search results list to select websites, and how to determine which listings are ads on a search engine results page). One also has to limit reliance on surface characteristics, such as the design of a website or a book (a website or book that appears to have a lot of information or looks aesthetically pleasant does not necessarily mean it has good information) and language used (a website or book that utilizes jargon, the keywords that one used to conduct the search, or the word "information" does not necessarily indicate it will have good information). The next phase is *information comprehension*, whereby one needs to have the ability to read, comprehend, and recall the information (including textual, numerical, and visual content) one has located from the books and/or online resources.

To assess the credibility of health information (*information assessment* phase), one needs to be able to evaluate information for accuracy, evaluate how current the information is (e.g., when a website was last updated or when a book was published), and evaluate the creators of the source—for example, examine site sponsors or type of sites (.com, .gov, .edu, or .org) or the author of a book (practicing doctor, a celebrity doctor, a patient of a specific disease, etc.) to determine the believability of the person/ organization providing the information. Such credibility perceptions tend to become generalized, so they must be frequently reexamined (e.g., the belief that a specific news agency always has credible health information needs continuous vetting). One also needs to evaluate the credibility of the medium (e.g., television, internet, radio, social media, and book) and evaluate—not just accept without questioning—others' claims regarding the validity of a site, book, or other specific source of information. At this stage, one has to "make sense of information gathered from diverse sources by identifying misconceptions, main and supporting ideas, conflicting information, point of view, and biases" (American Association of School Librarians [AASL], 2009, p. 13) and conclude which sources/ information are valid and accurate by using conscious strategies rather than simply using intuitive judgments or "rules of thumb." This phase is the most challenging segment of health information seeking and serves as a determinant of success (or lack thereof) in the information-seeking process. The following section on Sources of Health Information further explains this phase.

The fifth phase is *information management*, whereby one has to organize information that has been gathered in some manner to ensure easy retrieval and use in the future. The last phase is *information use*, in which one will synthesize information found across various resources, draw conclusions, and locate the answer to his or her original question and/or the content that fulfills the information need. This phase also often involves implementation, such as using the information to solve a health problem; make health-related decisions; identify and engage in behaviors that will help a person to avoid health risks; share the health information found with family members and friends who may benefit from it; and advocate more broadly for personal, family, or community health.

THE IMPORTANCE OF HEALTH LITERACY

The conception of health has moved from a passive view (someone is either well or ill) to one that is more active and process based (someone is working toward preventing or managing disease). Hence, the dominant focus has shifted from doctors and treatments to patients and prevention, resulting in the need to strengthen our ability and confidence (as patients and consumers of health care) to look for, assess, understand, manage, share, adapt, and use health-related information. An individual's health literacy level has been found to predict his or her health status better than age, race, educational attainment, employment status, and income level (National Network of Libraries of Medicine, 2013). Greater health literacy also enables individuals to better communicate with health care providers such as doctors, nutritionists, and therapists, as they can pose more relevant, informed, and useful questions to health care providers. Another added advantage of greater health literacy is better information-seeking skills, not only for health but also in other domains, such as completing assignments for school.

SOURCES OF HEALTH INFORMATION: THE GOOD, THE BAD, AND THE IN-BETWEEN

For generations, doctors, nurses, nutritionists, health coaches, and other health professionals have been the trusted sources of health information. Additionally, researchers have found that young adults, when they have health-related questions, typically turn to a family member who has had firsthand experience with a health condition because of their family member's close proximity and because of their past experience with, and trust in, this individual. Expertise should be a core consideration when consulting a person, website, or book for health information. The credentials and background of the person or author and conflicting interests of the author (and his or her organization) must be checked and validated to ensure the likely credibility of the health information they are conveying. While books often have implied credibility because of the peer-review process involved, self-publishing has challenged this credibility, so qualifications of book authors should also be verified. When it comes to health information, currency of the source must also be examined. When examining health information/studies presented, pay attention to the exhaustiveness of research methods utilized to offer recommendations or conclusions. Small and nondiverse sample size is often—but not always—an indication of reduced credibility. Studies that confuse correlation with causation is another potential issue to watch for. Information seekers must also pay attention to the sponsors of the research studies. For example, if a study is sponsored by manufacturers of drug Y and the study recommends that drug Y is the best treatment to manage or cure a disease, this may indicate a lack of objectivity on the part of the researchers.

The Internet is rapidly becoming one of the main sources of health information. Online forums, news agencies, personal blogs, social media sites, pharmacy sites, and celebrity "doctors" are all offering medical and health information targeted to various types of people in regard to all types of diseases and symptoms. There are professional journalists, citizen journalists, hoaxers, and people paid to write fake health news on various sites that may appear to have a legitimate domain name and may even have authors who claim to have professional credentials, such as an MD. All these sites *may* offer useful information or information that appears to be useful and relevant; however, much of the information may be debatable and may fall into gray areas that require readers to discern credibility, reliability, and biases.

While broad recognition and acceptance of certain media, institutions, and people often serve as the most popular determining factors to assess credibility of health information among young people, keep in mind that there are legitimate Internet sites, databases, and books that publish health information and serve as sources of health information for doctors, other health sites, and members of the public. For example, MedlinePlus (https://medlineplus.gov) has trusted sources on over 975 diseases and conditions and presents the information in easy-to-understand language.

The chart here presents factors to consider when assessing credibility of health information. However, keep in mind that these factors function only as a guide and require continuous updating to keep abreast with the changes in the landscape of health information, information sources, and technologies.

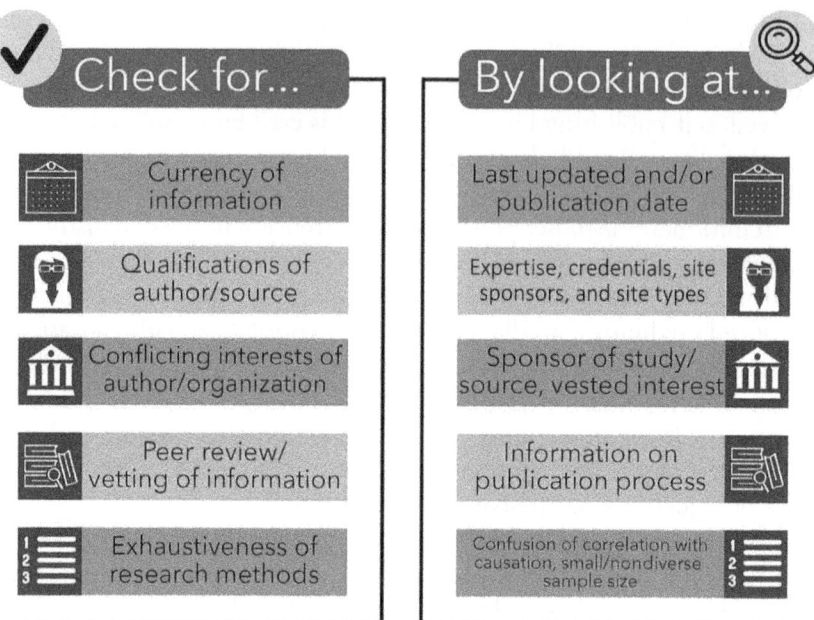

Check for...	By looking at...
Currency of information	Last updated and/or publication date
Qualifications of author/source	Expertise, credentials, site sponsors, and site types
Conflicting interests of author/organization	Sponsor of study/source, vested interest
Peer review/vetting of information	Information on publication process
Exhaustiveness of research methods	Confusion of correlation with causation, small/nondiverse sample size

The chart can serve as a guide; however, approaching a librarian about how one can go about assessing the credibility of both print and online health information is far more effective than using generic checklist-type tools. While librarians are not health experts, they can apply and teach patrons strategies to determine the credibility of health information.

With the prevalence of fake sites and fake resources that appear to be legitimate, it is important to use the following health information assessment tips to verify health information that one has obtained (St. Jean et al., 2015, p. 151):

- **Don't assume you are right**: Even when you feel very sure about an answer, keep in mind that the answer may not be correct, and it is important to conduct (further) searches to validate the information.
- **Don't assume you are wrong**: You may actually have correct information, even if the information you encounter does not match—that is, you may be right and the resources that you have found may contain false information.
- **Take an open approach**: Maintain a critical stance by not including your preexisting beliefs as keywords (or letting them influence your choice of keywords) in a search, as this may influence what it is possible to find out.

- **Verify, verify, and verify**: Information found, especially on the Internet, needs to be validated, no matter how the information appears on the site (i.e., regardless of the appearance of the site or the quantity of information that is included).

Health literacy comes with experience navigating health information. Professional sources of health information, such as doctors, health care providers, and health databases, are still the best, but one also has the power to search for health information and then verify it by consulting with these trusted sources and by using the health information assessment tips and guide shared previously.

Mega Subramaniam, PhD
Associate Professor, College of Information
Studies, University of Maryland

REFERENCES AND FURTHER READING

American Association of School Librarians (AASL). (2009). *Standards for the 21st-century learner in action*. Chicago, IL: American Association of School Librarians.

Hilligoss, B., & Rieh, S.-Y. (2008). Developing a unifying framework of credibility assessment: Construct, heuristics, and interaction in context. *Information Processing & Management*, 44(4), 1467–1484.

Kuhlthau, C.C. (1988). Developing a model of the library search process: Cognitive and affective aspects. *Reference Quarterly*, 28(2), 232–242.

National Network of Libraries of Medicine (NNLM). (2013). Health literacy. Bethesda, MD: National Network of Libraries of Medicine. Retrieved from nnlm.gov/outreach/consumer/hlthlit.html

Ratzan, S.C., & Parker, R.M. (2000). Introduction. In C.R. Selden, M. Zorn, S.C. Ratzan, & R.M. Parker (Eds.), *National Library of Medicine current bibliographies in medicine: Health literacy*. NLM Pub. No. CBM 2000-1. Bethesda, MD: National Institutes of Health, U.S. Department of Health and Human Services.

St. Jean, B., Taylor, N.G., Kodama, C., & Subramaniam, M. (February 2017). Assessing the health information source perceptions of tweens using card-sorting exercises. *Journal of Information Science*. Retrieved from http://journals.sagepub.com/doi/abs/10.1177/0165551 516687728

St. Jean, B., Subramaniam, M., Taylor, N.G., Follman, R., Kodama, C., & Casciotti, D. (2015). The influence of positive hypothesis testing on youths' online health-related information seeking. *New Library World*, *116*(3/4), 136–154.

Subramaniam, M., St. Jean, B., Taylor, N. G., Kodama, C., Follman, R., & Casciotti, D. (2015). Bit by bit: Using design-based research to improve the health literacy of adolescents. *JMIR Research Protocols*, *4*(2), paper e62. Retrieved from http://www.ncbi.nlm.nih.gov/pmc /articles/PMC4464334/

Valenza, J. (2016, November 26). Truth, truthiness, and triangulation: A news literacy toolkit for a "post-truth" world [Web log]. Retrieved from http://blogs.slj.com/neverendingsearch/2016/11/26/truth-truthi ness-triangulation-and-the-librarian-way-a-news-literacy-toolkit-for -a-post-truth-world/

Common Misconceptions about Essential Oils

1. ESSENTIAL OILS ARE CALLED "ESSENTIAL" BECAUSE THEY PLAY A SIGNIFICANT ROLE IN HUMAN HEALTH

The word "essential" actually refers to the oil that comes from the essence of the plant, an aroma-carrying substance that retains its scent when it is extracted from the plant itself. This oil lives in ducts or glands that are part of the plant's structure. In most cases, the oil evaporates when heated, so some users of these oils refer to them as "volatile" oils, a direct reference to this property of instant vaporization. Others call them "ethereal" oils, a term that gives them a mystical quality but that actually refers to their ability to disappear into the air (see question 2: What makes the oils "essential"?). A number of studies have suggested that certain essential oils can aid in sleep, reduce anxiety, relieve headaches, and reduce inflammation, but more research will be required to determine their actual effects in human beings. No study has proved that essential oils can be used to treat a serious illness.

2. ESSENTIAL OILS AND AROMATHERAPY ARE NOTHING MORE THAN NONSENSE AND PSEUDOSCIENCE

While the potential medicinal properties of essential oils are not fully understood at this time, their use for aromatherapy can have positive effects. Some studies have shown that people with chronic anxiety disorders, clinical depression, insomnia, low appetite, nausea, and even dry mouth can benefit from aromatherapy. The skin also can absorb the oils, so a massage or bath with a scented oil can help release tension, allowing a person to relax. This provides therapeutic benefits to someone with aching muscles or prolonged anxiety issues (see question 11: Do essential oils have medicinal properties?). The food manufacturing industry also is investigating the potential for essential oils' use as antimicrobials in packaging, in hopes of replacing many synthetic preservatives with more natural oils.

3. ESSENTIAL OILS ARE WELL REGULATED BY THE U.S. FOOD AND DRUG ADMINISTRATION

The U.S. Food and Drug Administration (FDA) does not regulate the processing, purity, procurement, bottling, or sale of essential oils. Essential oils are not cosmetics, foods, or drugs, so they do not fall under the jurisdiction of the FDA or any other U.S. government agency. Essential oils do come to the FDA's attention, however, when their sellers claim that these oils have the ability to cure disease. The FDA regulates all claims of therapeutic properties made by sellers in their marketing and on their packaging, to keep products that are not approved as drugs from being marketed as if they are drugs—especially when the claims are false (see question 5: How are essential oils regulated in the United States?).

4. ANY ESSENTIAL OIL IS SAFE TO USE INTERNALLY

Essential oils are plant based and are therefore considered natural products; many come from organically grown crops. The fact that they may be natural and organic, however, does not necessarily mean that they are safe to eat. The term "food grade" is used fairly freely by essential oils marketers, and many of these companies recommend the use of their oils in cooking. There is no official classification of "food grade" in the essential oil industry, however—no regulating agency defines what makes one oil edible or eligible to be a cooking ingredient, while another is not. This distinction becomes even vaguer in the packaged food industry, where many food products list essential oils in their ingredients (see question 33: Can essential oils be taken internally?).

5. ALL COMPANIES THAT SELL ESSENTIAL OILS CAN BE TRUSTED TO PROVIDE TRUE AND COMPLETE INFORMATION ABOUT THEIR ORIGINS AND USE

As in just about any industry, facts about essential oils can be buried under marketing terms, sales messages, and other fictions. Multilevel marketing companies—the ones that have lots of representatives that hold house parties to sell essential oils—have considerable history in making false claims about the medicinal and healing properties of their products. Purchasers of these products should be wary of any statement that the oils do anything more than provide pleasant scents and, perhaps, assist in relieving stress and a few minor ailments (see question 25: What is a multilevel marketing company, and should I buy essential oils from one?).

QUESTIONS AND ANSWERS

The Basics

1. What are essential oils?

Essential oils are called "essential" because they come from the essence of a plant. They are distillations of molecules called terpenes—also known as hydrocarbons—which provide a plant with its scent. Many essential oils also contain terpenoids, which are chemicals that also influence the scent, color, and flavor of the plant, and phenylpropanoids, which protect the plant from ultraviolet light and creatures that eat plants, and aid in the attraction of pollinators (bees and butterflies) by contributing to scent and pigment. The way a plant smells can help protect it from pests while attracting pollinators by releasing its scent into the air at the warmest time of day during the growing season. This is one of the reasons that flowers smell so good in spring and early summer. Some plants release their scent at night to attract moths and bats that also serve as pollinators.

Plants store their essential oil in structures throughout the plant, including glands, conduits, and cavities from which they can secrete the liquid in microscopic drops during pollination season. While we tend to think of plants producing scent within their flowers, the oil producing the aroma may actually be found in the stem, leaves, fruit, bark, and even the roots of the plant, creating quite a powerful system with which to attract bees, moths, hummingbirds, and other active pollinators.

These oils are gathered from a wide range of plants. More than 17,500 plant species produce an essential oil, though hardly more than 1 percent of these are harvested for their oil. The essential oils we can buy today come from plants that can be grown on farms—in other words, they thrive well in agricultural conditions where the potential for drought, insect infestations, and disease can be controlled. These plants also produce their oil fairly easily and dependably using one or more of the methods described later in this book (see question 4: How are essential oils obtained from plants?). Some plants do not surrender their oil using one of these methods, while others are too delicate to withstand the extraction process. Not all plants have a pleasant scent or taste, so the effort of cultivating them and extracting their oil may not be worthwhile.

Many purveyors of essential oils tout their medicinal properties beyond their scents. With more than 350,000 plant species identified around the world, however, only a tiny fraction has been assessed for their potential in healing or combating disease. Science has gravitated to the essential oils already on the market, as the claims of their usefulness as medicine have been passed down through more than a century of common usage. These familiar oils also have the advantage of being readily available for testing.

Today more than 100 essential oils are available commercially, obtained from showy flowering plants like rose, geranium, jasmine, and bergamot; from spice plants like cinnamon, ginger, and clove; from herbs like basil, oregano, and thyme; and from citrus fruits including orange, lemon, and tangerine.

Calling these substances "oils" is a little misleading, as they are not oils in the same way as canola, olive, or vegetable oil are. Oils used in cooking are known as fixed oils, because they do not change their state when heated. Canola oil, for example, continues to be a liquid even at high heat, because it contains fatty acids that prevent it from vaporizing.

Essential oils, however, are volatile oils, because they turn from liquid to gas as they get warm. This makes them especially appropriate for use in aromatherapy, as they can be warmed, vaporized, and inhaled easily.

Essential oils are obtained from plants using one of three methods: distillation, expression, or extraction, either using a solvent or a process called hypercritical CO_2 extraction. (See question 4: How are essential oils obtained from plants?)

2. What makes the oils "essential"?

These oils are called "essential" because they come from the essence (scent) of the plant, not because they are in any way critically important to human life.

While there is no way to know for certain who was the first to call these oils essential, the name appears to have had its origins in the Middle Ages, when alchemists (the precursors to pharmacists) distilled the oils from plants and vaporized them. These early scientists experimented with the oils in an effort to find ether, the elusive fifth element that they believed filled the empty space between the four tangible elements: earth, air, fire, and water. When their efforts in working with plants produced a scented vapor, they labeled this the "quintessence," seeing it as a first step in finding ether. Much more modern scientific research in 1887 dispelled the myth of ether, but the name for the easily vaporized scented liquids remained.

The term "essential" also implies that these oils excel in purity and that they have not been adulterated by chemicals. Products that contain alcohol, water, sugar, corn syrup, or other additives are labeled extracts (such as vanilla extract, one of the most common food flavor additives), perfumes, or flavorings, while the implication of essential oils is that they contain nothing but the essence of the plant in liquid form. Not all essential oils are 100 percent pure, however, despite labeling that advertises them as such. Some extraction processes require the use of solvents to obtain the terpenes from the plant, and these solvents remain with the oil as a result. Additionally, some bottlers of essential oils use additives to produce more oil using less of the expensive essence, resulting in greater profit for the bottler. This adulteration is not always apparent to the consumer.

3. Are essential oils 100 percent natural?

Some essential oils are exactly what they say they are: a pure distillation or expression from their plants of origin. Some, however, contain water, alcohol, or solvents used in their extraction (see question 4: How are essential oils obtained from plants?). Others are mixed with other substances derived from plants to extend the batch, allowing bottlers to stretch their supply of an expensive essence without the high cost of obtaining more of it. Some are synthetic, laboratory-produced versions that smell the same as the natural ones or that contain just the fragrance component of the more complex oil—and, therefore, may not be the natural product that consumers have been led to believe they are.

Labels that claim that an oil is "100 percent pure" may be honest or not, as no governing body or agency regulates these oils or determines what "pure" means for them. The U.S. Food and Drug Administration (FDA) prevents essential oil proprietors from claiming that these oils have specific medicinal properties or that they can protect users from viruses or bacteria that cause diseases. Under FDA rules, substances that make these claims must be laboratory tested and peer reviewed and will then be examined by the FDA, classified as drugs if the research bears out, and be federally regulated. Very little research has been done to prove that essential oils—individually or collectively—can guard against or cure any illness, so labeling and advertising generally do not make these claims. More to the point, the FDA plays no role in determining the purity of these oils or in substantiating any other claims made by companies marketing the oils.

It may seem on the surface that consumers have no way to tell which oils are pure and which are not, but there are clues to help buyers decide which companies may offer a purer oil.

First, essential oils should be packaged in dark brown or green glass bottles to protect them from vaporizing in bright light. Companies that package their oils in less costly clear glass know that their oils are in no danger of vaporizing or otherwise altering their chemistry, probably because they contain something more than 100 percent pure essential oil.

Every bottle of essential oil should display the plant's scientific or botanical name (in Latin) and the method used to extract it from the plant, as well as its point of origin—whether the crop from which this bottle was distilled came from lavender fields in New Mexico or the rosewood forests of Brazil. If the Latin name is missing, chances are that the content of the bottle is not strictly the essence of that plant. If there's no point of origin or method of extraction, the oil might have been assembled in a factory instead of being hand expressed in a barn halfway around the world.

Essential oils should be priced according to the availability of the oil and the complexity of the extraction process. Scarcer oils are priced higher, and essential oil sellers pass on the cost to the consumer of higher-priced processing for oils that are more difficult to obtain. Purveyors of these oils who price them all identically—for example, hawkers at carnivals and festivals whose vials of oils are all the same size, shape, and price—may have some adulterated oils in their inventory. If their oils also come in clear glass bottles, the alert consumer can be certain that the content of these bottles is not entirely what it seems.

Even the most well-known and trusted essential oils marketing companies, however, can sell oils that are not as pure as their labeling suggests. The only way to be certain of what is in an essential oil is to order gas chromatography-mass spectrometry (GC-MS) laboratory testing, a costly step most consumers will not take. GC testing separates the molecules of various substances from a gaseous sample to allow them to be identified and studied independently of one another. Mass spectrometry captures these compounds as the GC instrument releases them and sorts them according to their mass in less than a second, making it possible to identify and analyze them. These tests can identify if an oil is truly from the plant it says it is, as well as particulars like the geographic location of a particular cultivar or the oil's plant origin (such as the difference between lavender and lavandin, which come from the same family but have marked differences in scent, making one preferable to individual consumers over the other).

The process is not infallible, as many compounds are very similar and may be mistaken for one another, and some compounds in essential oils are still virtually unknown to modern science. Nonetheless, the analysis certainly will answer the question of whether or not the oil contains 100 percent of the compound(s) advertised or a number of different compounds that are not essential oils. GC-MS testing also identifies synthetic compounds in essential oils, making it particularly necessary and valuable for determining the oils' purity for their use as preservatives in organic foods.

4. How are essential oils obtained from plants?

Three basic methods are used to extract essential oils from plants: distillation, expression, and either solvent extraction or hypercritical CO_2 extraction. The most widely used method, involving both water and steam, dates back to 1928 when Joseph Franklin Clevenger published instructions for constructing an apparatus "for the determination of volatile oil," involving a flask to hold the plant material and some water, a separator to isolate the oil from the water in a graduated tube, and a condenser to gather the product of this process. Clevenger's steam distillation process replaced the previous, more cumbersome method of boiling the plant material in water and waiting until the plant's oil surfaced and formed its own layer on the water—a time-consuming endeavor at best. The Clevenger apparatus proved fragile, and correct placement of the separation valve became tricky, so in 1951, Jakub Deryng made modifications to the design and published his method. The Deryng design, with some modifications, remained in use into the twenty-first century, though much more modern techniques now include the use of microwaves and supercritical fluids—fluids like carbon dioxide and water, which have properties of both a liquid and a gas.

Distillation uses steam to deconstruct the plant and capture the vapor that rises from it. The vapor contains the terpenes than give the plant its scent. The plants stem, leaves, flowers, roots, or bark may be used in the process. Several different kinds of distillation are used for different plants, but they all involve placing the plant matter on a grid or screen in a sealed container so moisture can surround the material.

Water distillation places the plant directly in water inside the container and boils the water until it produces steam. The steam permeates the plant, causing it to fall apart and release the terpenes. This method is especially effective with flower blossoms, which are delicate and break down easily when steamed.

Water-and-steam distillation is less direct, placing the plant on the screen but keeping the water below it. The steam rises and penetrates the plant, capturing the terpenes as it elevates within the container.

Steam distillation eliminates the water inside the container; instead, the steam is injected into the container at high pressure from the bottom. As the steam rises, the plant comes apart, just as it does in the more direct versions of this process.

Hydrodiffusion distillation injects the steam from the top of the container instead of the bottom. This is particularly good for extracting the essence from tough parts of the plant like bark or a woody trunk.

In all these methods, the vapor that rises from the plant gets channeled into a condenser, where the steam cools. The resulting liquid contains water from the distillation process and the essential oil, which is lighter than water and floats on top of it. This makes it easy to capture by siphoning off the oil.

Oils collected from fruit rind or pulp are gathered using expression, sometimes described as cold-pressing. This mechanical process does not use heat but involves turning and puncturing the fruit in a machine and collecting the juice and essential oil beneath it. This allows the essential oil to rise to the top of the liquid, leaving the juice behind. Lemon, orange, lime, tangerine, and other oils pressed from fruits are obtained in this manner.

The third method, solvent extraction, is used when the plants themselves are too fragile to survive the steam pressure involved in distillation or the mangling that happens during expression. Fragile flowers like jasmine and gardenia would not yield their oils using these methods, so bottlers use solvents to extract the terpenes and chlorophyll from these plants, as well as some solid plant matter and wax that reside within the plant. This results in a compound called a concrete—having nothing to do with cement—which is then combined with alcohol to create an absolute, a scented material for specific uses like perfumes and cosmetics. Solvents involved in this method may include ethanol, methanol, hexane, or petroleum ether. These are removed at the end of the process, however, leaving just 5 to 10 parts per million (ppm) of residue from the solvent. Essential oils extracted in this manner are not appropriate for topical use or ingestion; they usually are used only for aromatherapy or as ingredients in nonedible, scented products.

Hypercritical CO_2 extraction uses carbon dioxide at very high pressure, transforming liquid into gas until there is no distinction between the two. The substance that results from this can pass through the plant matter and carry out the essential oil with it. This fairly new method has been in use in other industries, especially in removing caffeine from coffee beans.

5. How are essential oils regulated in the United States?

The short answer to this question is that essential oils are not regulated in the United States. As they are not food, drugs, or cosmetics, they do not fall under the purview of the FDA unless the companies that package and sell them make claims about their use that they cannot substantiate with scientific proof.

Essential oils fall into the same general category as dietary supplements, vitamins, and other substances that do not claim to have specific therapeutic value. There may be many claims about the oils' potential for healing and antimicrobial use in the world—most of it on the internet, provided by people and companies with no peer-reviewed scientific evidence—but the companies that produce these products cannot make these claims in their marketing materials.

This does not stop essential oils companies from claiming that the products are "therapeutic" and even inventing the terms "therapeutic grade," "Grade A," "medical grade," and "pharmaceutical grade," implying that some regulating body actually determines that some oils are more effective than others. No such agency exists to bestow such grades on any essential oil.

The FDA maintains a list of essential oils and other "natural extractives" that are Generally Regarded as Safe (GRAS) "for their intended use," which in this case is "food for human consumption." These include anise, basil, bergamot, black pepper, cassia, cinnamon, clary sage, clove, coriander, cumin, fennel, geranium, ginger, grapefruit, juniper berry, lavender, lemon, lemongrass, lime, marjoram, melissa, orange, oregano, peppermint, petitgrain, Roman chamomile, rosemary, spearmint, tangerine, thyme, wild orange, and ylang-ylang.

Even with the GRAS distinction, however, none of these oils should be consumed in quantity. One or two drops in a recipe will achieve the desired flavor; more than this will be overpowering.

Some essential oils are used as flavor additives by the food manufacturing and packaging industry. Like all natural and artificial flavorings, they are regulated by the FDA, and only those that make the GRAS list may be used in foods. Others have been banned worldwide by the International Fragrance Association in Geneva, Switzerland, for being fatal if swallowed, carcinogenic, or harmful if applied topically (see question 44: Are some essential oils toxic?).

With no official regulation of these oils and their purity, however, their use becomes a *caveat emptor* situation. The only entities between the oil and the user are the companies that market these oils, including the many representatives of multilevel marketing companies who are rewarded for higher sales regardless of the oils' usage or effectiveness.

Some essential oils' labels may state that the oils are "approved by the ISO." The International Standardization Organization (ISO) has compiled a list of botanical names of plants used in the production of 167 essential oils, as well as the English and French names for these oils, and rules for chromatographic and ester testing methods, preparation of test samples, packaging and storing the oils, and methods for assessing the purity of individual oils. When a package suggests that the oil is "approved" by the ISO, consumers may assume that it has met a standard for quality, but the standard actually has to do with the oil's chemical composition, not with a specific level of product quality.

Many of the claims made by unscrupulous marketers have been stricken from their websites and literature, because the FDA and the Federal Trade Commission (FTC) require that such claims be substantiated by research and approved for safety and effectiveness. When these companies made such claims without research to back them up, they were forced to walk back their hyperbolic advertising.

For example, if a company says on its website that one of its most popular essential oil blends can cure influenza, pneumonia, and cancer, this makes the product a drug according to the Federal Food, Drug, and Cosmetic (FD&C) Act. The FD&C Act defines drugs, in part, by their intended use: "articles intended for use in the diagnosis, cure, mitigation, treatment, or prevention of disease."

Based on the claims in its advertising, the company must receive approval by the FDA through its New Drug Application process to be classified as an over-the-counter drug. If the company has not applied for this status, the FDA will approach that company to inquire about the research it has conducted to back up its claim. If the company cannot produce research, the FDA will "advise" the company that it must remove the claims of therapeutic value from all of its marketing materials, usually within 48 hours, or face "legal action seeking a Federal District Court injunction." Such an injunction can be devastating to the company, leading to the products being recalled or pulled from the market entirely and the company refunding the purchase price to consumers. In some cases, such action could shut a company down.

Most recently, the FDA has sent warning letters to several essential oils companies to demand that they remove claims from their websites that their products can prevent or cure COVID-19. One such company promoted its oils with statements like this one: "The most important essential oils to protect against Corona Viruses are [list of oils] . . . This blend contains antiviral infection-fighting essential oils which are known to be effective in killing positive-stranded RNA viruses including Corona viruses." No peer-reviewed research suggests that this could be remotely true, so the FDA forced the company to remove these statements from its marketing communications.

As of this writing, no essential oils have been classified as over-the-counter drugs by the FDA for any purpose.

6. Historically, how were essential oils (or plant-based scents) used?

Plants were part of healers' earliest apothecary, probably long before the development of written language made it possible to catalogue each plant's purpose. As the practice of distillation did not emerge until it was discovered by Avicenna around 1000 CE, people collected plants with strong scents for a wide range of uses, from bringing comfort to the sick to masking the smell of a dead body until it could be buried. The Old Testament of the Judeo-Christian Bible provides instructions for using specific plants to consecrate temples and altars (see question 8: How are essential oils used in religious practice?), thousands of years before alchemists began distilling their essences.

Around 4500 BC, records indicate that ancient Egyptians soaked aromatic plants in oils (most likely olive, given the fruit's prevalence in the region) and used the resulting scented oil in ointments, salves, and poultices. Myrrh, grapes, onion, anise, and cedar all provided strong scents and apparent medicinal benefits, so these became staples in an Egyptian physician's apothecary. The first mentions of scented oils in Chinese and Indian medicine took place sometimes between 3000 and 2000 BC, with more than 700 substances mentioned in aged texts that survived to the modern era. This staggering list includes sandalwood, cinnamon, ginger, and myrrh, some of today's most popular essential oils.

Sometime around 1800 BC, a cadre of Roman and Greek physicians traveled to Egypt to learn from that country's renowned physicians, who not only practiced medicine but also recorded their successes and failures for the common good. One of the tools these early doctors espoused was the use of plant material and essences for specific ailments, some mixed with carrier oils or used in making preparations to apply to patients' bodies. The Greek and Roman physicians brought this information back to their own territories, and the oils became part of the Greek medical tradition. Greek physicians embraced cumin, marjoram, peppermint, saffron, and thyme for their perceived medical properties, as their records noted for the first time between 500 and 400 BC. The Romans, however, lost or forgot their own knowledge of scented plants and their use in healing when the Roman empire fell between 376 and 476 CE.

Centuries passed before the knowledge of plant essences and their ability to heal would re-emerge, spurred forward by Persian-born Avicenna's method for capturing plant scents and flavors in liquid form through distillation. The Arabs took the next step, becoming the first to ferment sugar and distill ethyl alcohol from it. This new substance turned out to be a better solvent for extracting plant essences than oils were, making the essences more accessible to alchemists in the 1200s. By this time, Marco Polo made his famed journey to Asia and brought back spices including cardamon, cinnamon, nutmeg, and sage, to the delight of merchants—and to physicians, who began to experiment to determine if these new spices and herbs had any use in treating disease and wounds.

Medicine moved slowly in the age before modern laboratories, so the next solid mention of precursors to essential oils came in the 1500s in Europe, when the list of oils included benzoin, calamus, cedarwood, cinnamon, costus, myrrh, rose, rosemary, sage, and turpentine. A Swiss physician, Paracelsus, became a thought leader in the use of these oils for medicinal purposes, pushing other physicians and alchemists to work to extract plant essences from leaves, roots, and wood or bark as well as from stalks.

By the 1700s, as many as 100 different essential oils were now part of a pharmacist's stock. What medicinal value these oils actually have still remains to be seen three centuries later, but chemists in many different fields undertook the task of determining each oil's chemical composition. This opened the door for the use of these oils in a wide range of applications, from perfume to food flavoring.

In 1937, French chemist Rene Gattefosse published *Aromatherapy*, a book about the use of inhalation of the vapor from scented oils to treat diseases. His evidence of this technique's effectiveness turned out to be largely anecdotal or based on a small number of cases, but just as books touting miraculous quick-fix cures top the bestseller lists today, Gattefosse's theories gained traction and popularity. Aromatherapy remained a European phenomenon until the early 1980s, when Jean Valnet, a French physician, published his own book, *The Practice of Aromatherapy*. This one crossed the ocean and received attention in the United States, refreshing and advancing the essential oils industry here.

The twentieth century saw another platform emerge: essential oils as home remedies for everything from hangnail to asthma, as well as their use in cleaning products. When aromatherapy became popular, essential oils entered the mainstream to become a multi-billion-dollar business with many growers, bottlers, packagers, and marketing companies making them easily available to the general public.

7. Who uses essential oils today, and how common is essential oil usage in the United States and around the world?

Essential oils have found a significant place in a number of industries, including food and beverage, personal care and cosmetics, and aromatherapy. Some early research suggests that selected oils may find their way into pharmaceutical and medical applications, according to a May 2020 market analysis report by Grand View Research, a global market research firm. While the global market for the oils is valued at a staggering US$18.62 billion in 2020, growth over the next seven years is expected to nearly double that figure, to $33.26 billion. Europe holds the largest share of the market with more than 43 percent of oils sold there.

The mainstream medical community in the United States has not embraced essential oils as an alternative therapy for any specific health condition, but the general public has seen a significant increase in their usage for aromatherapy and perceived health benefits. As stress levels rose throughout the 2020–2021 pandemic, some consumers sought alternatives to doctor visits, turning to essential oils as a readily available stress reliever with the perceived ability to help users become more focused on tasks and more mindful and "in the moment" while working at home. A diffusion of essential oils could make working from home feel less challenging and a little more like a benefit; if it helped the home office worker feel less stressed and more centered, so much the better.

The rise in the number of day spas across the United States has been a significant contributor to the increased demand for essential oils. Luxurious scents play an important role in helping the spa visitor feel relaxed and pampered, so spa proprietors make liberal use of essential oils to provide clear, clean scents that smell more natural than commercial air freshener aerosols. Beauty salons that offer waxing for hair removal may use essential oils to cool inflamed brows and upper lips while scenting the entire establishment with one or more oils in a diffuser.

The growth of health beverages including kefir, kombucha, and energy drinks has bolstered the essential oils market as well. In an effort to appeal to health-conscious consumers, manufacturers have turned to essential oils as flavorings in place of laboratory-made flavored compounds or sugary juices.

The oils are readily available to consumers through many retail channels—displays of the oils now appear in supermarkets, big-box retail stores, and many drugstores—but the largest sales channel is direct selling through multilevel marketing companies, including Young Living Essential Oils and doTerra. These companies do much of their business through gatherings in the homes of their salespeople or in the homes of friends who agree to host parties to help them sell. These companies market the oils individually and in blends, with statements that suggest the potential for each blend to boost mood, heighten alertness, or even engender feelings of self-worth, spiritual and self-awareness, reverence, connection, relaxation, strength, power, balance, well-being, comfort, and awareness of one's unlimited potential.

Online sales have allowed smaller companies with specialty lines to reach consumers looking for specific oils or blends, as well as more exotic oils than they can obtain from the larger companies.

On the industrial side, essential oils are used in many packaged foods, especially in baked goods and other sweets. Interest in the oils and the antimicrobial properties touted by some marketing companies has led to increased scientific research to determine if these claims have some basis in fact. The research, sponsored by food packaging companies, seeks an organic alternative to preservatives now used in many foods. If some of the essential oils do indeed have the ability to ward off fungus and mold, they could become critically important to organic food companies—and the pharmaceutical industry may take an interest as well.

While Young Living and doTerra dominate the multilevel marketing sector, many companies have sprung up as the demand for essential oils grows. The global company Research & Markets predicts that the essential oil market may be on the verge of a series of mergers and acquisitions, in an effort to narrow the field of small players marketing their own lines of essential oils.

8. How are essential oils used in religious practice?

All of the world's major religions have used natural scents in their worship, in ceremonial rites, and in daily practice to aid in creating an environment that opens the mind to celebration of their deity. Scented oils and plants often play an important role in the oldest scriptures.

In the Old Testament of the Judeo-Christian Bible, for example, God directs Moses and Aaron to use myrrh, cinnamon, and calamus to make holy anointing oil, which they would use to consecrate the "ark of the testimony" in the first temple (Exodus 30:23). Jews use myrtle, citrus, palm, and willow as the Four Plants in the waving ceremony throughout the seven-day feast of *Sukkot*, or harvest; citron (*etrog*) and myrtle (*hadass*) provide the pleasant smells that symbolize the doing of good deeds, while willow (*aravah*) has no taste or smell, signifying people who do no good deeds and do not study the Torah. Palm (*lulav*) has an appealing taste but no smell, so it represents people who study the Torah but do not follow through with good deeds. Myrtle and citrus are now available as essential oils, though the waving of the Four Plants still requires the use of fronds or small branches from each.

Of the many Judeo-Christian scriptural references to trees, spices, and herbs that modern chemists now turn into essential oils, the best known comes from Matthew (2:11) in the New Testament, the gifts that the three wise men bestow on to the baby Jesus and his parents in Bethlehem: frankincense and myrrh, prized in biblical times for their disinfectant and anti-inflammatory properties. The gift comes full circle at the end of Jesus's life, when Nicodemus brings more than 100 pounds of myrrh and aloes with which to bind "the body of Jesus . . . in strips of linen with the spices, as the custom of the Jews is to bury" (John 19:39–41). Today scented incense is part of many Catholic rituals, and Christians and Jews can incorporate the essential oils extracted from the 30 plants mentioned in the Bible into their personal religious practice.

Many religions use incense in their rites of worship. Buddhist temples honor their ancestors with it and use many different kinds of resins and woods in their practice to clear the mind for meditation. Flowers and incense are part of Buddhist *pūjās*, rituals done before a statue of the

Buddha, to show respect and honor to the Buddha's memory. The Buddha himself, however, emphasized virtue over scents, saying in verses 54 and 55 of the Dhammapada, the Buddha's Path of Wisdom, "Not the sweet smell of flowers, not even the fragrance of sandal, *tagara*, or jasmine blows against the wind. But the fragrance of the virtuous blows against the wind. Truly the virtuous man pervades all directions with the fragrance of his virtue. Of all the fragrances—sandal, *tagara*, blue lotus and jasmine—the fragrance of virtue is the sweetest." In China, Buddhists use camphor in the construction of rosaries, a powerful scent that awakens and relaxes. Juniper berries, pine, and sandalwood are also popular and traditional.

The use of incense, oils, and other scented materials has been a part of Chinese religious culture since ancient times. *Gāoxiāng*, the word meaning "high incense," signifies the use of incense as a religious offering to ancestors or in worship of a deity. Some Sunni Muslims in China also use incense as a part of worship, while others denounce this as a holdover from other religions, including Buddhism.

Hindus burn incense during their worship as a symbol of their aspiration into the heavens, as well as an homage to Agni, the god of fire referenced in the Rigveda sacred text. In Vedic sacrifices—a number of different rituals performed with the goal of achieving discipline and raising consciousness to the heavens—aromatic substances were often included in the offerings to God.

A number of oils appear in Islamic texts including the Hadith. Before prayer, which they undertake five times daily, Muslims wash their hands, feet, and other body parts to remove any impurities before they handle a holy book or enter a mosque. They may add essential oils to the water before they bathe, perhaps choosing camphor, as instructed to do for burials and funerals by the messenger of Allah in Hadith 1458: "Wash her three or five times, or more than that if you think you need to, with water and lote leaves, and put camphor or a little camphor in (the water) for the last washing. When you have finished, call for me." Hadith 4346 places worshippers who are of lower status on "sandhills of musk and camphor, and they will not feel that those who are sitting on chairs are seated better than them."

◆❖◆

Uses and Benefits of Essential Oils

9. What does scientific research tell us about essential oils?

Until recently, the global scientific community had not taken a great deal of interest in essential oils. In the past 20 years, research into the biological activities of essential oils has increased, providing the first glimmers of insight into the oils' potential medicinal properties. Laboratory research has revealed that essential oils have a complex chemical composition, making it difficult to isolate a single pathway through which they affect microbes. Their complexity, however, may be exactly what makes some of these oils effective against bacteria, viral cells, or fungal cells in a test tube. Some oils are producing positive results in laboratory tests, so research will continue to determine if the results produced in a lab can be replicated in human beings.

This does not mean that claims of health benefits made by marketers of essential oils are true or that any essential oil can serve as a substitute for an antibiotic, chemotherapy, or other modern medical therapy recommended by doctors.

No statement can be made about the properties of essential oils as a whole, because each has its own unique chemical composition. Over the past several decades, studies of 10 specific oils have yielded interesting information, some of which may lead to the use of one or more of these oils as part of a pharmaceutical solution—but whether these oils will turn out to be alternative antibiotics, pain relievers, anxiety reducers, sleep aids, or the cure for cancer is unknown at this time.

What is known, however, is that essential oils are potent and should be used sparingly. Because of the way they are distilled or expressed from tons of plant matter, the oils are 50 to 100 times more concentrated than they are in the plant. They should not be applied directly to skin without diluting them in a carrier oil (see question 31: What are carrier oils, and why should I use them?), because the potency of some oils can irritate the skin or even cause chemical burns.

Scientists warn that consumers should not ingest the essential oils they purchase in stores or from multilevel marketing companies, even though some of these received the Generally Regarded as Safe (GRAS) ranking from the U.S. Food and Drug Administration (FDA). Oils on the GRAS list have been approved for use in food manufacturing, where minute amounts of them are blended into massive quantities of ingredients; even a drop or two in a glass of water may be too strong for home use. In addition, aromatherapy diffusers should not be left on all day—a 30-minute time period is long enough to fill a room with scent, without overexposing adults, children, and pets to the oil. No essential oil should be used on young children.

Research has shown that aromatherapy has some positive effects on the human body and psyche. Scientists theorize that the components of essential oils bind to receptors in the brain's emotional center, triggering the limbic system and producing a calming or invigorating effect. This has been observed through functional imaging studies, in which researchers monitor brain activity using computerized tomography or magnetic resonance imaging while the subject inhales air containing the vaporized essential oil. The studies have also noted that aromatherapy has very low toxicity, producing little to no harmful vapor—and diluting the oils in a carrier oil to apply to the skin also reduces the oil's potential toxicity.

Other recent studies have examined essential oils' antimicrobial and other potential medicinal properties; we'll discuss these in turn later in this book.

10. How is the research limited in what it can tell us about essential oils and their possible benefits?

Scientific research struggles with testing essential oils for a range of reasons.

First, no generalized statement can be made about the efficacy of the oils in combatting any microbe, because there are well over 100 different oils on the market today. Essential oils are distilled or expressed from flowers, leaves, stems, bark, trunks, and resins, each with its own chemical composition. Even the oils from a specific plant can be different: for example, oil distilled from lavender is not the same as oil distilled from lavandin, as these are two distinct plant varieties from one botanical family.

Second, oils labeled from the same plant of origin may come from different geographic areas. Lavender gathered from a farm in Albuquerque, New Mexico, may not have the same properties as lavender grown in Provence, France. Differences in soil composition, fertilizer, and agricultural techniques may change the chemical structure of the oil.

The complexity of each oil's composition also creates challenges for laboratory research. One bottle of lavender essential oil may have a different chemical makeup from another, even if they come from the same supplier. Plants gathered and processed in the morning may be altered somewhat when compared to plants harvested the night before. The components will be the same, but they may vary in concentration.

The oils' effectiveness often varies from one study to the next, because the oil used in each study may be markedly different. With no federal regulation of what actually goes into each bottle, scientists struggle to be sure that the oil they test in a laboratory in California is the same as the one tested in India. Replicating results from one study to the next becomes an obstacle to success.

Researchers have also found it particularly challenging to conduct blinded studies with scented substances. Scents trigger emotions and memories, so if one group is tested using lavender and the control group is tested with a different scent, the second scent may inadvertently jostle a memory loose or create an emotional response in a control group subject. If the study is about cognitive ability or emotional reaction to a scent, the random triggered memory may distort the results of the study. If, on the other hand, the study is meant to test the physical therapeutic value of the lavender, the control group may be tested using an alternative scent that is known to have no positive or negative effect on the body. This may or may not lead to a calculable outcome.

To date, most studies have been phase I: in vitro (in a test tube), determining an oil's effect on microbes or tissue; or phase II: on laboratory animals. A few studies have involved human subjects, but these have only a few participants, not the large populations involved in true phase III, double-blind testing of a drug or vaccine. These studies are too small to produce statistically significant results, so their findings are considered interesting but anecdotal. Currently, the best research involves reviews of many small studies, bringing together all the studies around the world and performing a meta-analysis of the results to draw conclusions. Such papers often reveal correlations that isolated studies cannot see.

A 2019 review of many studies, published in the June issue of the peer-reviewed journal *Molecules*, determined that while many essential oils do demonstrate antimicrobial properties against a wide range of diseases and infections, the success of one variety of an essential oil cannot be generalized to others of the same species. There are more than 400 species of thyme, for example, and while some of them have performed well in a test tube (not yet in humans) against viruses like herpes simplex and influenza H1N1, as well as both Gram-positive and Gram-negative bacteria, only specific thyme species have this capability. Such results could be used by unscrupulous sales channels to dupe the public into thinking that their thyme oil—whether or not it is the correct species—will help them ward off these viruses and bacteria. It will be some time before there is enough research to determine the correct species and the industry's ability to produce large quantities of it, as well as the dosage and method of use that will make any specific essential oil marketable as a pharmaceutical.

This brings us to the biggest challenge involved in scientific research on essential oils: someone has to pay for the research, and the usual channels—pharmaceutical companies—have not taken a great deal of interest in doing so. As essential oils are natural substances, they cannot be patented as stand-alone drugs. This limits the ability of Big Pharma to make a profit with them in their pure form. If the limited research in progress does produce a positive result, pharmaceutical companies may eventually find ways to combine the oil with other products to create a patentable drug. This may accelerate the pace of continued study.

Essential oils have been in use for thousands of years, so researchers often want to take a shorter route to producing test results that reveal their effects on humans—that is, skipping the in vitro and animal testing steps and going right to testing them on humans. This makes traditional funding sources like the National Institutes of Health balk, as they prefer to fund research that follows the standard guidelines regardless of the substance being tested. Proposed studies are abandoned if the researchers can't get the funds they need to move them forward, so essential oils languish, becoming the neglected stepchildren of medical research.

11. Do essential oils have medicinal properties?

It is difficult to say for certain that essential oils as a whole can have an effect on disease, illness, pain, or any other medical condition, as there is a great deal of variation between oils and their effects on the human body. Given the long history of plant matter in ancient medical practice, folk medicine, and alternative therapies, however, much research currently explores this question.

Ancient practitioners of healing and medicine thousands of years ago used plant essences to treat all kinds of ailments and diseases, with varying degrees of success. Aromatic oils appeared in medical practice in Egypt, China, and India as far back as 2,500 years ago, but whether those patients' health simply improved with time and rest or because of the use of a specific oil can never be known for certain.

Hippocrates (460–370 BC), considered an early founder of Western medicine, believed strongly in the healing power of nature (*vis medicatrix naturae* in his writings) and often used ointments and balms made from plants to treat wounds and other injuries. (Aromatherapy practitioners are fond of quoting him as writing, "The key to good health rests on having a daily aromatic bath and scented massage," but an exhaustive search of his writings did not produce this statement and included only a passing reference or two to aromatics.)

Modern medicine requires rigorous testing of any substance to determine its effectiveness in treating a specific condition. Until very recently, the claims made about the healing properties of essential oils had not attracted much attention from the scientific community, so very little testing had been done in the United States and only slightly more overseas. There has been a significant increase in testing since 2010, and some of this research has produced promising results in the laboratory, with mixed results in human trials. The results of these tests may lead to more consideration of essential oils' potential as medicine for human beings.

Each oil has its own chemical composition, so with dozens of these oils available, no general statement can be made about all of them—just as we cannot state that all prescription pharmaceuticals have an identical effect on the human body. We do know, however, that essential oils are not required for human health and are not called "essential" because of any role they can or do play in human existence. The term simply refers to their chemical nature within the plants from which they are extracted.

Harpreet Gujral, DNP, FNP-BC, program director of integrative medicine at Sibley Memorial Hospital in Washington, DC, notes that even if certain scents do nothing but elevate a person's mood, they can have a positive effect on their health. Inhaling a pleasant scent sends molecules through the olfactory system to the amygdala, the part of the brain that governs emotions. If the aroma pleases the person smelling it, they may feel better emotionally, which can improve their physical health.

Today's professional aromatherapists point to a wide range of properties attributed to essential oils. Some are considered analgesics (painkillers), some may have antidepressant capabilities, and others are considered antifungal, anti-inflammatory, antiseptic, antispasmodic, calming, decongesting, aid to digestion, disinfectant, fever-reducing, sedative, diuretic, or a booster to the immune system. We will look at many of these uses and the truth or fiction in these claims later in this book.

Specific oils have shown some promise in scientific studies. For example, doctors have used peppermint oil to calm gastrointestinal distress for many decades, and recent meta-analysis (Alammar, Wang, et al., 2019) of 12 randomized trials of a total of 835 patients has shown that it has a positive impact on irritable bowel syndrome (IBS) as well. A 2017 study published in the *American Journal of Gastroenterology* (Chowdhary et al., 2017) found that peppermint oil relieved swallowing disorders (dysphagia) in two-thirds of the patients who used it.

Tea tree oil has become an ingredient in a number of products for topical application. One randomized, double-blind, placebo-controlled study as far back as 2007 determined that a 5 percent tea tree oil gel had as much effect on reducing mild to moderate acne as benzoyl peroxide, the leading over-the-counter pharmaceutical for the same purpose (Enshaieh et al., 2007).

Lavender—or more specifically, linalool, one of the terpenes in the plant's oil—has been shown in studies to have a calming effect on mice with a sense of smell, in comparison to mice with no ability to smell the oil (Harada et al., 2018). To further test this effect, the researchers injected the mice with flumazenil, a drug used by anesthesiologists to reverse the effects of sedatives, and the mice responded as they would if they had taken the sedatives. This told Harada and his team that linalool had had its effect on the same receptors in the body that are calmed by sedatives, making the terpene a potential calming agent on its own. Baldwin and Chea (2018) conducted a similar experiment on eight dressage horses and found that aromatherapy using lavender calmed the heart rate variability in the horses as well, while chamomile did not—indicating that the lavender, not floral scents in general, could be used to lower anxiety in naturally skittish show horses.

A mist of water scented with lavender essential oil (specifically *Lavandula angustifolia*) has been shown to have an effect on sleep, helping people fall asleep more rapidly and stay asleep through the night. The impact of the scent on rodents in the study (Buchbauer et al., 1993) was described as "significant," putting the mice to sleep fairly quickly. In a second round of testing, the researchers stimulated the mice to the point of "overagitation" with caffeine and then exposed them to lavender again, and the mice became sleepy even though chemicals in their system pushed them to stay awake.

Tea tree oil and eucalyptus oil have each shown promise in studies examining their use as antiviral treatments for herpes simplex virus (HSV). Schnitzler (2019) reviewed all the studies to date on essential oils and HSV and determined that these two appear to have the ability to inhibit HSV from replicating or attaching itself to cells (adsorption). These laboratory results need to be tested in clinical trials with humans, but these oils in pharmaceutical form may be a solution to an uncomfortable and recurring virus.

More research must be done before any of these potential uses can be put into practice. Simply picking up a bottle of tea tree oil and dabbing it into a herpes lesion will not cure the virus; factors including dosage, form and method of use, frequency, safety, and other issues must be determined before essential oils can be used in a medical application.

12. Should I believe claims that essential oils can cure diseases?

Statements about essential oils curing any kind of medical condition must be met with skepticism until the scientific community has completed its research. The important thing to remember is to consider the source of the information: no one who will profit from the sale of the oils should be giving out medical advice. This takes us back to the snake oil salespeople of earlier times: they had no proof of their claims, but they managed to convince thousands of people to drink their magic elixirs and, when their symptoms cleared up by coincidence, actually believe that they were being helped by the swill.

People who have health issues often will search for solutions beyond whatever they have been offered by doctors, especially if they are fighting a chronic condition or a life-threatening one. Some forms of alternative medicine have proven benefits, while others do not—and for the most part, studies of essential oils have not yet provided definitive proof that they are effective in fighting illnesses or diseases. Joy Victory, deputy managing editor of HealthNewsReview.org, noted in a 2017 article that essential oils are "being marketed to tense, health-conscious Americans searching for inexpensive 'natural' cures." Sellers of these oils are "trained to carefully make generic health statements—e.g. 'It supports healthy digestion!'—to escape the attention of federal regulators who would flag more specific claims."

As essential oils gained in popularity in the late 1990s and early 2000s, one marketing company claimed to have found evidence that a band of robbers in the fifteenth century tied bundles of specific aromatic plants into scarves that they wore over their noses and mouths, to protect them as they looted the homes of people who had died of one of the era's contagious diseases. The oils in these plants, the story goes, kept the men from contracting the pathogen and getting sick, allowing them to become quite wealthy during the pandemic. If these thieves avoided the plague, the company's advertising claimed, it stands to reason that our oils will keep you from getting sick.

We who have survived the 2020–2021 pandemic know that wearing a double cloth face mask would have been good enough to keep the vast majority of people from getting the virus; the plants the fifteenth-century thieves wore (if they wore them at all) would have provided nothing more than a little extra filtering material. The story continues to be told, however, and while the FDA ordered the company to take down claims displayed on their website that their oils can prevent modern-day viruses, representatives of the company speaking one-on-one with customers may be quick to point to a special blend of oils that they market as the one true antiviral concoction.

There's another culprit in leading consumers to believe that essential oils have proven health benefits: the popular media. Stories in many media outlets including *USA Today, Prevention, Women's Health,* many wire services, and news websites provide the points of view of people who sell essential oils, claiming that they "may" treat all manner of health conditions, from weight loss to muscle pain to IBS. Some of these articles go so far as to cite a preliminary study, but these are often pre-publication studies that have not been reviewed by a peer group of scientists to determine if their research methods are sound. Worse, most of the spokespeople quoted have a vested interest in selling essential oils. These stories do not go so far as to locate scientific researchers who can provide real information about their own studies—instead, they focus on the claims of salespeople. This makes any statements in the articles suspect.

For any claim of medical benefit, the information should come from a credible source: a peer-reviewed scientific journal, for example. Even pre-publication websites that contain scientific papers can fall into the questionable category, as the researchers' methodology and conclusions may not measure up to the rigorous standards followed by most scientific journals. The National Cancer Institute's page on aromatherapy and essential oils recommends that a reputable study describe clinical findings "in enough detail that a meaningful evaluation can be made" and that it report on a therapeutic outcome—a "measured improvement in quality of life," for example. If, on the other hand, the information is anecdotal, the conclusions will benefit a specific company, or if there are no details about how, when, or where the study was conducted, it should not be taken as medical advice.

13. Should I use essential oils instead of prescribed medications?

Essential oils are meant to enhance your quality of life by adding pleasing scents to your environment, and they may provide the added benefit of an elevation of mood, a calming influence, or even a remedy for low-level symptoms like clogged sinuses or headache. Some have additional benefits that are still being studied through science, like disinfecting surfaces or relieving joint pain. No research to date has revealed them as curatives for serious illness.

If you are considering ignoring a doctor's advice and attempting to treat an illness or disease with essential oils instead of medication, ask yourself these questions:

- What dosage should I use?
- How should it be administered?
- How often should I use it?
- How will I know if it's working?
- How much time should I give the essential oil to work?

Prescription or over-the-counter medications always come with instructions for their use, as well as lists of potential side effects and the length of time you should take the medication to be sure it has done its job. As essential oils are not medications, they do not come with specific instructions about how much to use to treat an illness or symptom. No medical professional has determined that X number of drops in Y carrier oil applied to Z place on the body will knock out a patient's sinus infection or shrink a cancerous tumor. If you don't know what dosage is purported to work, how often you should use it, and in what way it should be administered (aromatherapy, topical, or ingested), you have no way to know if it's going to work or not. This is not how medication is meant to function.

Let's say you have made the admittedly ridiculous decision to treat a tumor with essential oils instead of chemotherapy and radiation. How much oil should you use, where on your body should you apply it, and how will you know if it is doing the job? You cannot see the tumor, so it may be getting larger as you spend valuable time trying to treat it with an unproven therapy. Without computerized tomography or magnetic resonance imaging, you will not be able to tell if the tumor has shrunk or grown in the time that you have spent attempting to treat it with oils. Will you return to your physician after rejecting her recommendation to use traditional or leading-edge medical therapies and tell her what you have been doing instead?

Without specific instructions about the use of one or more essential oils, there is no way to know how long you should dose yourself with the oils before you can hope to see an improvement. What if you begin to feel worse? If you have put your faith entirely in the oils instead of proven medical therapies, you may wait and wait for some evidence that they have started to work, and you may wait much too long, endangering your own life in the process.

Of course, essential oils can't shrink a cancerous tumor, so you would never attempt to treat one with them. If you have chosen to treat any illness or condition with the oils instead of following your doctor's instructions, however, you put yourself at risk, as well as your family if the illness is contagious. Essential oils are no substitute for Western medicine. Use them to ease symptoms like anxiety or headache or as a complementary therapy for more serious issues, but always follow your doctor's orders.

14. What is an antimicrobial essential oil, and which essential oils have this property?

An antimicrobial agent has the ability to stop or slow the growth of microorganisms such as bacteria, fungi, and viruses or to kill them altogether. The term "antimicrobial" is an umbrella category for antibacterial (antibiotic), antifungal, antiseptic, and antiviral agents.

Some studies have shown that a few essential oils have "demonstrated potential" as antimicrobial agents, but so far these results have emerged only in laboratory studies, in which the essential oil reacts against the microbe in a test tube. Lavender essential oil, for example, had a "strong antiseptic effect" against antibiotic-resistant strains of *Staphylococcus aureus* (also known as MRSA) in a laboratory study in 2012. A study that tested 13 essential oils to determine their effect on major respiratory tract pathogens found that cinnamon and thyme oils demonstrated the strongest action against the microbes, with clove essential oil as the next strongest. In both these cases, phase II studies using animals (usually mice) and phase III studies in human subjects would be required before scientists can declare any of these oils to be effective for people who actually have one of these diseases.

Eucalyptus has long been used in treating respiratory tract infections such as bronchitis and sinusitis, for its ability to open congested nasal and bronchial passages with its cooling vapors. Can it actually cure such an illness? A study published in the June 2012 issue of *BMC Complementary Alternative Medicine* explored this question, testing eight species of eucalyptus essential oils harvested from Tunisia, to see if any of them had an effect on the microbes that cause influenza, pneumonia, bronchitis, and other respiratory infections. In a laboratory study, one of the species, *E. odorata*, "showed the strongest activity" against several of the pathogens, while another one, *E. bicostata*, seemed to have an effect against viruses. However, the researchers noted that the activity "diminished with the decreasing essential oil concentration." As in many of the studies that have shown positive results, the essential oils produce a much weaker and less robust antimicrobial effect than synthetic pharmaceuticals produce, making them significantly less effective even in a test tube than current prescription medications.

As essential oils can reach the respiratory tract easily through inhalation, logic would suggest that respiratory infections might be the best place to start to determine if the oils in vaporous form can affect various kinds of illnesses. In July 2018, a study published in *BMC Complementary Alternative Medicine* explored this theory, evaluating the effects of clove, cinnamon bark, eucalyptus, thyme, scots pine, peppermint, and citronella essential oils against the bacteria that cause pneumonia, various illnesses in the *Streptococcus* family, two *Haemophilus* influenzae, and one of the bacteria that causes respiratory and middle ear infections. They found that thyme had some effect against *S. mutans*, and cinnamon bark and clove oils showed "high inhibition" against pneumonia and one of the forms of strep. Cinnamon bark, in fact, turned out to be the most effective of all the oils tested on the widest range of pathogens. The researchers concluded that clove, cinnamon, and thyme "may provide promising antibacterial activity" against pathogens found in the respiratory tract, but "their effect is lower than that of the reference antibiotics." They posited that the essential oils might be used in combination with antibiotics, but more research would be required to determine a recommended dose and whether taking enough of the essential oils to have an effect could turn out to be toxic.

By now you may start to see the challenges involved in researching whether or not essential oils can actually be effective in fighting viruses and infections. Phase I results produced in a test tube against an isolated virus or bacterium are only the beginning of the research cycle; such studies cannot by definition reveal the effects of any substance on an actual human being. Even if the phase I (laboratory) results look promising, they must be very strong to convince a funding organization such as the National Institutes of Health to provide a grant to take the research to the next level. If the results are not triumphantly positive—or at least as strong as the existing pharmaceuticals currently in use—most research ends there.

15. What is an antibacterial essential oil, and which essential oils have this property?

Bacteria are one-celled organisms that cannot be seen without a microscope, living just about everywhere inside and outside the human body. Many bacteria are good for us, including those that live in the digestive tract and break down the foods we eat into nutrients and waste. Others cause diseases, illnesses, and infections. Tuberculosis, strep throat, staph infections, urinary tract infection, cholera, pneumonia, diphtheria, botulism, typhoid, meningitis, tetanus, Lyme disease, and many venereal diseases are all caused by bacteria.

Purveyors of essential oils have long claimed that many of these oils have the ability to guard against harmful bacteria. Until very recently, no scientific evidence existed to suggest that these claims might contain a kernel of truth. Thanks to the interest of the food packaging industry, however, a number of studies now point to the potential of essential oils as antibacterial agents, perhaps making them valuable in preventing bacteria from developing in packaged meats and other perishable products.

A study published in *Microbios* (Pattnaik et al., 1996) found that eucalyptus, lemongrass, orange, and peppermint oils were effective against 22 bacterial strains, while aegle (Indian bael) and palmarosa oils inhibited 21 different kinds of bacteria. The following year, Pattnaik broke down the oils into their constituents to determine what characteristic of the oils had the greatest effect on bacteria. He discovered that linalool, one of the aromatics in the oils, inhibited 17 of the 18 bacteria on which it was tested, while cineole and geraniol prevented the growth of 16 of the bacteria.

A 2016 study by Radaelli et al. published in the *Brazilian Journal of Microbiology* examined the antibacterial activity of six essential oils commonly used as condiments in Brazilian cooking. The results indicated that rosemary, basil, marjoram, thyme, and peppermint essential oils all had the ability to kill bacteria in a laboratory, while anise, the sixth oil tested, did not. "The use of essential oils from these common spices might serve as an alternative to the use of chemical preservatives in the control and inactivation of pathogens in commercially produced food systems," the paper concluded.

More recently, in 2017, Puskárová et al. tested six essential oils— arborvitae, clary sage, clove, lavender, oregano, and thyme—to determine their antibacterial properties against a range of bacteria including *E. coli*, salmonella, staphylococcus, listeria, and others. The team found that oregano, thyme, clove, and arborvitae performed well against all the bacteria tested, even when the oils were used in reduced concentrations rather than at full strength. The study experimented with two application methods: researchers applied the oils directly to the bacteria in a laboratory and also infused them as vapor. The vapor trail determined that the oils inhibited the bacteria as well as the direct-application method did, apparently indicating that they could be used to kill bacteria in an open environment, such as disinfecting the air in a room.

A study published in October 2020 in the journal *Antibiotics* tested 15 essential oils in a laboratory setting against strains of *E. coli* and *S. aureus* (MRSA) that are proving to be resistant to traditional antibiotic pharmaceuticals. The researchers broke the oils down into their constituents and tested each of them individually against the targeted bacteria. They found that compounds called thymol and carvacrol, found in *Lippia origanoides* essential oil—an oil already in use as a food preservative—showed the most antibacterial activity of all the substances tested. They cautioned, however, that this oil has the potential to be toxic in the human body, so it would not necessarily be a viable alternative for patients with MRSA or *E. coli* to take internally.

These are just a few of the studies that indicate that some essential oils have the ability to fight harmful bacteria. A review of all the research on essential oils through 2016, published in the journal *Evidence-Based Complementary Alternative Medicine* (Swarmy et al., 2016), noted that three plants in the *Achillea* (yarrow) family have shown an ability to counteract certain pneumonia bacteria, as well as *S. aureus*, *E. coli*, and strep and staph germs. Six varieties of *Artemisia* (daisy) appear to have bacteria-fighting components, as do specific species of ajwain, anise, basil, bay laurel, bitter melon, black pepper, cinnamon, citronella, clove, coriander, cumin, cypress, fennel, kumquat, lavandin, juniper, menthol, melaleuca (tea tree), myrtle, oregano, parsley, patchouli, rosemary, summer savory, *Salvia* mints, thyme, verbena, and *Warionia*, a Saharan plant. This long list of plants and their constituents have been the focus of much scientific research worldwide over the past 20 years.

All the research has taken place in laboratories using isolated compounds and bacteria in agar or Petrie dishes, however, and only a few studies have reached phase II, in which the substances are tested using laboratory mice. To date, no studies have produced recommendations for use of these oils to defeat infections in the human body—which means that there are no guidelines to suggest how much of the oil would be required to treat a person, how it should be administered, and if there could be side effects from the oil's potential toxicity.

A day may come when essential oils can be used to cure human illness, but that day is not here yet. Much more research will be required to translate the early findings of these studies into practical use to benefit actual patients.

16. What is an antifungal essential oil, and which essential oils have this property?

Fungi are single-celled or multicelled organisms that must get their nutrients from other organic material.

- Some fungi live in water (Chytridiomycota), where they can give fungal infections to frogs and other aquatic creatures.
- Some eat decaying plants or animals, becoming the mold that grows on human food (Zygomycota). Mold on bread and baked goods comes from this family of fungi, as does the molds used to make many kinds of cheeses.
- Fungi in the Ascomycota family, including many kinds of yeast, are particularly toxic to humans and can cause conditions including athlete's foot, ergotism (which can lead to hallucinations and convulsions), and ringworm. Some of these yeasts live in the human body, including *Candida albicans*, which can be triggered into excessive growth that causes yeast infections and candidiasis—a potentially fatal infection.
- Other fungi provide benefits to humans and plants. Glomeromycota gain nutrition from plants and transform plant sugar into minerals that they deposit into the earth, thus nurturing the plants from which they get their food.
- Basidiomycota include mushrooms and truffles—some of which are quite edible and beneficial, while others are poisonous.

The medical community has seen a recent rise in the number of fungal infections in patients, in large part because of systemic conditions that suppress the immune system. New drug therapies often treat the illness without curbing the accompanying fungal infection, and doctors sometimes do not detect the infection until it becomes life threatening. Fungal infections can affect agriculture as well, with significant losses of entire crops to microbial diseases—for example, the green mold *Fusarium graminearum* attacked the U.S. wheat crop between 1998 and 2000, causing an estimated loss of $2.7 billion to the wheat industry.

With so much loss of life and livelihood, experts in the fields of medicine and agriculture have spent considerable time searching for solutions to the fungus problem—one that will not cause further harm to the patient or spread toxic pesticides on cropland. One potential answer may be forthcoming through research on the antifungal effects of some essential oils.

Researchers have selected several oils on which to focus their efforts: thyme essential oil for its concentrations of thymol and carvacrol, two known antimicrobial components; tea tree oil, loaded with terpenes that kill fungi; peppermint essential oil; and clove essential oil. Another cluster of studies looked at the benefits of myrrh essential oil against topical fungi on the skin.

Pattnaik's study in 1996 (see question 15: What is an antibacterial essential oil, and which essential oils have this property?) confirmed that aegle, citronella, geranium, lemongrass, orange, palmarosa, and patchouli essential oils all inhibited the growth of 12 different fungi, while eucalyptus and peppermint oils discouraged 11 of these fungi. When he broke the oils down into their constituents in a subsequent study, Pattnaik found that citral and geraniol were the specific factors in the essential oils that prevented the fungi's growth. Linalool was nearly as effective as citral and geraniol, while cineole and menthol inhibited just 7 of the 12 fungi.

A study published in the *Journal of Medical Microbiology* (Pinto et al., 2006) examined the effects of a specific thyme essential oil, *Thymus pulegioides* (lemon thyme or broad-leafed thyme), on fungi including several strains of Candida, aspergillus (which causes the lung infection aspergillosis), and dermatophytes, fungi that cause infections of the skin, scalp, and nails. The thyme oil showed "significant activity" against the fungi in a laboratory, "deserving further investigation for clinical applications." A 2010 study by Vale-Silva et al., published in the German journal *Planta Medica*, found even stronger results in applying thyme essential oil (*T. viciosoi*) to several fungi in a lab, observing "rapid metabolic arrest, disruption of the plasma membrane and consequently cell death." This ability to break down the cell's outer coating and destroy the cell appears to make thyme essential oil particularly effective in neutralizing fungal infections.

Calendula essential oil also appears to have strong antifungal properties, according to several studies. The *Brazilian Journal of Microbiology* published a study (Gazim et al., 2008) that found *Calendula officinalis* to have strong potential antifungal activity when tested against 23 fungi. A study published in the *Journal of Environmental Science and Health* (Císarová et al., 2016) concluded that clove, thyme, and oregano essential oils had the strongest effect on *Aspergillus* fungi of the 15 essential oils tested, though "all essential oils exhibited activity against all tested strains of fungi."

The journal *Scientia Pharmaceutica* published a study in August 2020 that tested thyme essential oil from the *T. vulgaris* plant species to determine its effectiveness in treating topical inflammation and fungal infections. Boukhatem et al. (2020) found what others had discovered before them: a "potent anti-inflammatory effect at all doses." Just two months later, in October, *Frontiers in Cellular and Infection Microbiology* published a paper (Parrish et al., 2020) in which the researchers detailed their study of 65 essential oils and 21 essential oil blends against a range of dermatophyte strains. Twenty-one days after treating the dermatophyte spores with the oils, the team found that cassia, cilantro, cinnamon, thyme, and oregano were the most effective against the fungi, as was doTerra's proprietary DDR Prime blend, which contains clove, thyme, litsea, and wild orange essential oils.

A study published in *Pharmaceutical Biology* (Mahboubi and Kashani, 2016) analyzed the ability of myrrh extract and myrrh essential oil against dermatophytes in a laboratory. The researchers found that myrrh essential oil was more effective than the extract against topical fungus, leading them to conclude that their study "confirmed the traditional uses of C. molmol [myrrh] as a poultice for the treatment of cutaneous fungal infections."

All these studies end with a call for more research. No pharmaceutical products have emerged to treat fungal infections with thyme essential oil, however, as the required phase II (mouse models) and phase III (human) studies have not been completed; moreover, until their publication, we have no way to know if they are even in progress. The common wisdom among companies that market essential oils is that thyme oil can remedy toenail fungus, one of the most stubborn conditions to treat, so the oil and various blends are available for this purpose. Some homeopathic products touted to treat topical fungi contain calendula essential oil, another one of the several that seem to be potent against these conditions.

17. What is an antioxidant essential oil, and which essential oils have this property?

When the human body encounters harmful substances in its environment—cigarette smoke, vapor from electronic cigarettes, ozone, industrial chemicals, pesticides, fried foods, ultraviolet light, and others—it produces molecules known as free radicals. These unstable molecules contain an unpaired electron, so their entire existence revolves around finding an electron with which to bind, allowing them to stabilize. This process creates a condition known as oxidative stress, because oxygen is required for the bonding to occur. Each time a free radical binds with another molecule, it steals an electron from that molecule, which creates another free radical that must find a replacement electron. This process has the potential to damage proteins, cell membranes, and even our DNA, the building blocks that determine every characteristic in the body—and when DNA becomes damaged, it can mutate into the kinds of cells that cause cancerous tumors to grow. Free radicals also can hasten the aging process, and they can encourage the development of degenerative diseases including heart disease, multiple sclerosis, Parkinson's disease, autoimmune diseases, and dementia. It is critically important for the body to maintain a balance of free radicals and antioxidants, nutrients that keep free radicals from forming and reduce their damage to the body.

Antioxidants come from phytochemicals, substances found in plants, so eating fruits and vegetables plays a key role in controlling free radicals. Taking antioxidant supplements does not have the same effect, however—in fact, some studies suggest that synthetic antioxidant supplements may actually do harm to the body. Research tells us that eating fresh fruits and vegetables is required to gain their positive antioxidant effects.

How do essential oils fit into this picture? Essential oils come from plants, making them candidates for antioxidative properties, but to date, only a few have been studied for their potential ability to counteract free radicals. A Russian study published in 2009, for example, studied 14 essential oils using gas-liquid chromatography to observe their antioxidant behavior in a laboratory. The study determined that garlic, clove bud, ginger, and cinnamon leaf essential oils had "maximal efficiency" of 80 to 93 percent in preventing oxidation. These oils have not been studied for their effects on oxidative stress in the human body, however, so no conclusions can be drawn to suggest that they would be useful as antioxidants for people.

A 2010 study published in the *Journal of Agricultural and Food Chemistry* took the research to the next step, testing coriander and caraway essential oils in laboratory animals. Samojlik et al. (2010) pretreated the specimens with either coriander or caraway essential oil and then gave them carbon tetrachloride, which has a negative effect on the liver and kidneys. The researchers found that caraway essential oil "strongly inhibited" liver damage through its antioxidative ability, while coriander actually had the opposite effect—it acted as a prooxidant, encouraging oxidative stress. "The essential oil of caraway appeared promising for safe use in folk medicine and the pharmaceutical and food industries," the paper concluded.

Another 2010 study published in the journal *Natural Product Research* tested six of the most popular essential oils—lavender, peppermint, rosemary, lemon, grapefruit, and frankincense—for their potential as scavengers of free radicals. Yang et al. (2010) determined that lavender and lemon essential oils had the strongest antioxidant properties specifically against the free radical compound DPPH, and peppermint essential oil foiled the ABTS free radical. Lavender oil had the most lasting effect, continuing to fight free radicals 10 days after the test began.

As with most studies of essential oils and their potential effects on human diseases and conditions, the research cited here provides only a hint of the information required to put essential oils to use as antioxidants. Whether these oils can have a positive effect on oxidative stress through inhalation, consumption, topical application, or some other method remains to be seen, as there have been no studies to date involving human subjects.

18. What is an astringent essential oil, and which essential oils have this property?

An oil, lotion, cream, or other substance is considered astringent if it contracts the skin cells when applied topically. This contraction tightens pores to reduce the amount of natural oil on the skin, making astringents particularly desirable for teenagers fighting acne during puberty and adolescence. Astringents also can help reduce bleeding from a minor cut or scrape.

Most commercially available astringent products contain isopropyl alcohol, making them particularly strong and potentially irritating to the skin. Alcohol-based astringents can dry the skin to the point of itching and flaking, which can make skin appear dull and damaged. Using an appropriate essential oil as an astringent eliminates the medicinal scent of alcohol, a desirable improvement for those who use astringents on a daily basis.

The scientific community has not seen much of a need to research the astringent properties of essential oils, so most information on this comes from anecdotal evidence—but these anecdotes have been borne out by hundreds of years of evidence. We can take for granted, for example, that lemon essential oil provides an astringent effect when applied to the skin, leading to its use in many skin care products for oily complexions. Applying lemon oil directly to the skin can cause redness and peeling, however, so it must be mixed with a lotion or carrier oil that will not clog the pores (a "noncomedogenic" oil). Mixing a drop or two of lemon essential oil into a face washing product can be enough to reduce blemishes over time. Other citrus essential oils including grapefruit and bergamot are touted to be effective astringents as well.

Peppermint and spearmint essential oils, also well known for their ability to control oily skin, contain significant amounts of menthol, which has the added beneficial effect of cooling redness and irritation. While several oils that come from conifers (cone-bearing trees) have astringent properties, cypress essential oil is most often recommended as a remedy for oily skin.

Many practitioners who make their own astringent lotions recommend mixing a few drops of the selected essential oil with distilled water and apple cider vinegar, which contains its own astringent acids. Some also add rosewater—not rose essential oil, a very expensive product; rosewater is made by boiling water and adding rose petals and steeping this until the water turns pink. Witch hazel, also a powerful astringent, appears in some recipes. With the vinegar, witch hazel, and rosewater involved, you hardly need the essential oil—but a few drops will provide an additional astringent effect.

If acne has already made its appearance, the scientific community agrees that tea tree essential oil can be as effective in fighting pimples as benzoyl peroxide, the active ingredient in many over-the-counter acne products. Tea tree is not an astringent, but its antifungal and antibacterial properties make it a top choice for treating the inevitable results of oily skin.

19. What is a febrifuge essential oil, and which essential oils have this property?

A febrifuge is a substance that reduces fever. Fever occurs when the hypothalamus, an area of the brain that controls body temperature, shifts your normal temperature to a higher point. The human body's average normal temperature is 98.6°F (37°C), though your personal normal temperature may be slightly above or below this point. Your temperature may become elevated when something is happening to your body that does not normally take place, such as a viral or bacterial infection, or when you have been exposed to heat for a long time (a condition called heat exhaustion). Some inflammatory conditions like rheumatoid arthritis also can elevate body temperature. Many people develop a fever after receiving a vaccine, as the body develops antibodies to counteract the immunization; this is normal and will pass in a day or so.

Symptoms of fever include chills and shivering, alternating with sweating, headache, muscle aches, dehydration, dizziness, loss of appetite, and a general feeling of weakness or malaise. Generally, a fever of 102°F or less is simply an indication that you are fighting an illness, and it will pass in a short time (hours or a few days). If a fever rises above 103°F (39.4°C), it is time to call a doctor.

Commonly used and highly effective fever reducers in our daily lives include aspirin, acetaminophen, and ibuprofen, over-the-counter pain relievers that can also lower body temperature caused by an illness. A few essential oils can provide cooling relief during a fever when applied to the back of the neck, the forehead, or the soles of the feet, though none of these are as effective as aspirin for actually lowering body temperature.

Peppermint, spearmint, and eucalyptus essential oils contain menthol, a natural coolant that can be soothing when rubbed on the neck, chest, or back. Because menthol has the ability to be absorbed through the skin, it provides some penetrating relief that may actually reduce a fever slightly.

A more effective way to use essential oils to reduce fever may be to add them to a cold pack to place on the patient's forehead. Make a cold pack by filling a glass or metal bowl with about a pint of water and adding three or four drops of peppermint, spearmint, or eucalyptus essential oil. Fold a hand towel or washcloth to a size that will fit comfortably on a forehead, and place it on the surface of the water in the bowl. Let it sink into the water until it is wet through, then lift it out, wring out the excess, and wrap it once in plastic wrap to keep it from dripping. Place it on the patient's forehead, and let it rest there until it warms to room temperature. If nothing else, this should feel calming and refreshing to the person with the fever; it also may help to lower their body temperature.

Science has not bothered much with testing essential oils for their ability to reduce a fever, for the simple reason that we already have a solid solution for this: the over-the-counter medications mentioned earlier. Practitioners of alternative medicine do note that other essential oils have febrifuge potential, including basil, bergamot, black pepper, ginger, lemon, and patchouli. The oils that contain menthol are likely to provide the most benefit, however, because of their ability to penetrate beyond the skin.

20. Which essential oils are good for wound care?

A deep wound requires a doctor's care, but everyday cuts, scrapes, and bruises may benefit from the use of specific essential oils that have the ability to reduce the possibility of infection, heal damaged skin, and calm pain. As some kinds of infection become increasingly resistant to anti-biotic pharmaceuticals, alternative therapies like essential oils may gain interest and popularity.

Wound healing happens in four overlapping phases, as detailed by Barreto et al. in a 2014 paper published in the journal *Molecules*: coagulation (bleeding cessation), inflammation (pain and discoloration), formation of new tissue, and remodeling, which determines "the strength and appearance of the healed tissue." Barreto's team examined the literature on the effectiveness of monoterpenes, chemical compounds found in nearly all essential oils, on the various phases of the healing process. They found that every study that evaluated monoterpenes found them to be effective in wound healing, no matter what plant or oil they came from.

Research on the effects of essential oils in wound care is largely absent from the scientific literature, but a few studies reveal that some of these oils may have legitimate healing properties. Lavender and bay laurel essential oils emerged as the most active wound healers in a study using laboratory animals, published in the *Journal of Essential Oil Research* (Süntar et al., 2014). Just a few months earlier, in December, the *Journal of Alternative and Complementary Medicine* published a study using human subjects (Chin and Cordell, 2013), in which patients who had wounds with active *S. aureus* infections were treated with dressings that contained tea tree essential oil. "The results demonstrated decreased healing time in all but one of the participants treated with tea tree oil," the report concluded. "The results of this small investigational study indicate that additional study is warranted."

In August 2020, the *Journal of Alternative and Complementary Medicine* published a review conducted at the University of Nebraska of all the relevant studies using lavender essential oil in wound care. The researchers found 36 studies and included 20 of them in their screening: seven human trials, five animal trials, two in vitro studies, and six reviews of additional literature. Rachel Samuelson et al. (2020) determined that the studies indicated that lavender essential oil did indeed demonstrate "a faster rate of wound healing, increased expression of collagen, and enhanced activity of proteins involved in the tissue remodeling process" when the wounds were treated with lavender oil. They cautioned, however, that a number of different varieties of lavender essential oil were used in the various studies and that the chemical composition of the oil needed to be standardized to determine which was most effective. The researchers also recommended additional clinical trials with human subjects.

A single case detailed in a paper published in December 2009 in the German journal *Forschende Komplementärmedizin* involved a 41-year-old woman who developed a "minor, non-bleeding lesion" on her foot while gardening. The tiny wound became infected within hours, and the infection spread beneath her skin nearly to her ankle. Doctors prescribed oral antibiotics, but these were ineffective against the bacteria, and the wound eventually required surgery to remove the abscess. Five days after the surgery, the patient's doctor began a course of essential oils applied topically, choosing several oils "according to their anti-inflammatory, analgesic, and antimicrobial properties." The paper credits the oils for the wound's eventual healing, noting that the oils seemed to encourage rapid growth of healthy skin without further complications. With surgery in the mix, however, it is difficult to give all the credit for healing the wound to the essential oils, making this study a prime example of the use of a single instance—essentially an anecdote—to declare something true.

There are a surprising number of such papers among the scientific literature. For example, a paper published in January 2019 in the *Journal of Pediatric Nursing* examined the comparative case study of two children who had been severely burned. The grandmother of one of the children treated one child's burns with essential oils (the specific oils are not named in the paper), while the other child received standard medical treatment for burns. Jopke et al. (2019) noted that the child who received standard care developed two bloodstream infections and four "hospital-acquired conditions," while the child whose wounds were treated with essential oils developed just one acquired condition and stayed in the hospital for four fewer days than the other child. "While these case findings are intriguing, research is needed to expand understanding of the role of essential oils in the treatment of burns," the paper concludes.

Some essential oils are considered styptic, meaning that they can cause a wound to stop bleeding. Applied topically, a styptic agent, also called an antihemorrhagic or a hemostatic agent, can be used in an emergency to slow and stop the flow of blood from a cut, scrape, or larger wound.

Geranium, benzoin, bergamot, and yarrow essential oils are said to have this property, although there are no peer-reviewed studies to confirm that this is the case.

21. What is a cephalic essential oil, and which essential oils have this property?

A cephalic substance helps to clear the mind, allowing users to focus their attention on a task or subject. Some essential oils used in aromatherapy are believed to have this ability and may provide a bit of a stimulant as well—which, in turn, leads to better focus. Practitioners of essential oils claim that basil (ocimum), clary sage, garlic, juniper berry, peppermint, and rosemary oils can act as cephalic agents.

Anecdotal evidence abounds for the effects of essential oils on alertness and focus, but there is precious little science to back this up. Several studies examine whether inhaling the aroma of an essential oil can ward off sleepiness—not quite the same thing as promoting alertness but a step on the way to improved focus and concentration.

In 2005, Norrish and Dwyer's work with peppermint oil, published in the *International Journal of Psychophysiology*, pointed to a benefit from aromatherapy. The researchers introduced the scent of peppermint oil in a darkened room and placed their subjects in the room for 11 minutes. They used the Stanford Sleepiness Scale to determine the participants' overall sleepiness, as well as to observe changes in the activity of their pupils. "When compared with a no-odour condition, the presence of peppermint oil limited the increase in sleepiness during 11 min spent in a darkened room," they observed. "However, the mechanisms by which peppermint oil has its effect and the applicability of these findings to situations in everyday life will require further empirical investigation."

A recent study published in the May 2021 *Complementary Therapies in Clinical Practice* tested rosemary essential oil on 80 nurses working the overnight shift in a hospital. After administering two sleep questionnaires to the nurses to create an objective scoring system of their overall sleepiness, the researchers placed one drop of rosemary essential oil in the masks of 40 of the nurses and one drop of distilled water in the masks of the others. Later in their shift, the nurses took the sleepiness surveys again. The 40 nurses who received the rosemary oil saw their sleepiness scores drop significantly, indicating that they felt much less sleepy throughout their shift. The nurses who received distilled water, however, actually felt sleepier by the end of their shift. "Rosemary aroma decreased sleepiness and increased alertness in shift-working nurses," the researchers concluded.

A 2013 study in Austria, published in *Scientia Pharmaceutica*, used an electroencephalogram (EEG) with 20 human participants who had inhaled the aroma of rosemary essential oil. Sayorwan et al. (2013) found that inhaling rosemary oil reduced alpha waves in the brain, indicating a period of relaxation while awake, and it increased beta waves—the waves that are most active when a person is focused, alert, and ready to make decisions.

Sowndhararajan et al. (2016) conducted a similar study to determine the effects of *Inula helenium* (known colloquially as elecampane) essential oil on brain activity, using an EEG to monitor the participants' brains as they inhaled the oil's aroma. Their study, published in the *European Journal of Medicine* in August 2016, found "significant changes" in brain activity, with increases in some ratios between brain waves that signaled that *I. helenium* "may enhance the alertness state of the brain and could be used for the treatment of psychophysiological disorders."

A 2018 study of 79 high school students in Ukraine divided them into three groups: a control group, a group in a room sprayed with lavender essential oil, and a final group in a classroom sprayed with rosemary essential oil. Filiptsova et al. (2018), who published in the *Alexandria Journal of Medicine*, found that the group that inhaled rosemary essential oil had "significantly increased . . . image memory compared to the control," especially in the memorization of numbers. The lavender group had a "weakened" ability to memorize numbers, while the control group performed the worst.

So there seems to be some credence to the claim that rosemary and elecampane essential oils can foster a greater sense of alertness and focus, while peppermint essential oil may be able to banish the sleepiness many of us face in the middle of the day or when we don't get enough rest. Other oils may have this ability as well, though they have not been studied empirically, so any evidence of their cephalic qualities remains anecdotal at best.

22. What other properties are attributed to essential oils, and is there science to back up these claims?

Companies that sell essential oils tout a wide range of ills that these substances can affect, though most of these claims have not been tested using the scientific method. A basic online search reveals anecdotes, declarations, and wild assertions that essential oils can cure anything from a headache to a compromised immune system. Some users insist that essential oils cured their cancer, epilepsy, and even COVID-19. Whatever ailment you wish to correct, someone online will be happy to tell you that you can do it with a magical combination of essential oils.

These claims do not have much of a basis in science at this time, though essential oils have gained considerable attention from the clinical community in the past decade. A report to the U.S. Veterans Health Administration (VHA; Freeman et al., 2019) examined 26 systematic reviews of aromatherapy and essential oils for a wide range of health conditions. This exploration resulted in "moderate-confidence evidence" that aromatherapy may benefit women who experience menstrual pain or painful contractions during childbirth. The report goes on to indicate that essential oils may be useful in reducing blood pressure, alleviating stress and anxiety, lifting mood, and aiding sleep quality, with "low to moderate confidence in the evidence." For all other conditions, however, the report concludes that "there is insufficient evidence of efficacy."

Indeed, a randomized, double-blind study conducted by (Ou et al. 2012) used massage with a blend of lavender, clary sage, and marjoram essential oils to reduce the pain of menstrual cramps and found that the group that received the aromatic massage reported a shorter pain duration, from 2.4 days to 1.8 days. Another study (Bahr et al., 2018) used massage with doTerra's Deep Blue essential oil blend and the company's Copaiba oil to relieve the pain and stiffness of rheumatoid arthritis in patients' hands and found that participants who received the essential oil massage could complete tasks requiring dexterity 50 percent faster than those who received the placebo. The people whose massage contained the oils also experienced "increased finger strength, and significantly increased angle of maximum flexion compared to subjects treated with coconut oil," the study concluded.

A lengthy review of a wide range of essential oils for their individual constituents' effect on pain receptors, published in the journal *Molecules* (Sarmento-Neto et al., 2016), determined that a number of these oils contain specific chemicals that reduced pain in laboratory rats or mice. The results suggested that some of these constituents could be harvested from the essential oils for use in new pharmaceuticals, some of which might eventually replace opioids and other addictive drugs currently used to treat pain. Oils in specific combinations seemed more effective than individual oils on their own, the researchers noted. "Data from this review show the potential of essential oils like low cost analgesic drugs for new treatments for pain," they concluded. "However, there are few reports of toxicity to confirm therapeutic safety and to carry out clinical trials."

A study published in the *Journal of Clinical Gastroenterology* (Khanna, MacDonald, and Levesque, 2014) assessed randomized, placebo-controlled trials involving a total of 726 patients to test the use of peppermint oil for IBS, a condition with few medical treatment options. The review discovered that peppermint oil was superior to a placebo for treating the symptoms of IBS, although it did cause some heartburn in many patients. "Peppermint oil is a safe and effective short-term treatment for IBS," the researchers concluded. "Future studies should assess the long-term efficacy and safety of peppermint oil and its efficacy relative to other IBS treatments including antidepressants and antispasmodic drugs."

So essential oils may turn out to be effective against many medical issues, if enough funding emerges to pay for studies that can reach conclusions with clinical trials. For every study that suggests that essential oils may have therapeutic value, however, there are many claims with no scientific basis at all. Websites that claim that "[e]ssential oils cure everything!" rarely have a basis in fact, especially if they are written by people with no medical background.

One of the most potentially damaging claims online is that some combination of essential oils might guard against or even cure COVID-19. These claims are not based in science. No studies have shown that essential oils have an effect against this coronavirus or its variants—the few literature reviews completed by medical researchers have simply suggested that some essential oils might have a limited effect on symptoms. To date, no studies have proceeded past the laboratory (phase I) stage—there have been no studies with lab animals and none involving clinical trials with humans. Do not be misled by claims that essential oils will protect you against this potentially deadly virus.

Buying Essential Oils

23. How are essential oils sold?

For much of their recent history, essential oils have been sold by multilevel marketing (MLM) companies, most famously by Young Living Essential Oils and doTerra. Other MLM companies that sell essential oils include Arbonne International, Modere, Pruvit Ventures, and Total Life Changes. MLM companies make each sale of the oils a one-on-one experience, with a salesperson (called a "representative") explaining the use of each oil, the techniques for blending them with carrier oils, and the touted benefits of each. This process takes place at house parties, in spas and beauty salons, and in other places where customers—generally women—come together on a regular basis (see question 25: What is a multilevel marketing company, and should I buy essential oils from one?).

Many essential oil packagers and sellers also offer "collections" or "starter kits" that contain a selection of oils considered fundamental to a user's needs. These usually include 8 or 10 oils used commonly for diffusion, topical application, or in blending products at home, like cleaning solutions or laundry detergent. Some of these kits also provide a bottle or two of the manufacturer's proprietary blends with names that imply that their use will help the consumer release stress, reduce respiratory congestion, increase energy, and more. The kits can be pricey, depending on the oils and other products (such as hand cream or a roll-on bottle) included, with some approaching $200 at 2021 rates. Before purchasing a starter kit or a collection for a specific purpose, it would be better to choose one or two common oils—lavender and lemon, for example—and see if they truly become part of your new daily regimen before spending a great deal of money on others.

With their increase in popularity in the 2000s, however, essential oils have become more readily accessible than in any other time in their history. The oils can be found in supermarkets in the skin care or natural remedies aisle, in big-box discount stores like Target and Walmart, in specialty houseware stores, and in most national drug store chains. Most of these stores carry a range of individual oils and a selection of blends— essential oils combined by the manufacturer with an eye toward simplifying the purchase decision for consumers. Blends—whether sold in stores or by MLM reps—usually have names that denote the manufacturer's claim of benefit: Brave, Calm, Anchor, Forgive, Awaken, Believe, Motivation, Release, Sleep, and so on.

The global shift to online sales in the 2000s and 2010s has opened up essential oils to a massive consumer market, providing an interested user with the ability to purchase any commercially available oil they choose—all without tracking down a representative and navigating the time-consuming in-person sales process. Now if someone reads that the Judeo-Christian God directed Moses to mix several ingredients including galbanum, a fairly rare oil, to consecrate the tabernacle that would contain the Ten Commandments, they can search for galbanum online, compare prices and claims of purity, and own a bottle within days. This has made it possible for smaller or more specialized essential oils companies to develop a presence in the marketplace. Los Poblanos Organic Farm in Albuquerque, New Mexico, for example, grows its own organic lavender and makes a range of skin care products using the essential oil; thanks to its online marketplace, the little farm in the New Mexico foothills has developed a national audience for its oil and the products that contain it.

Generally, essential oils are sold in glass bottles, each containing about half an ounce (or less) of the oil. This may sound like not nearly enough to be effective, but most diffusers only require a drop or two of an oil to fill a room with fragrance, so a single bottle can last for months or even years.

Essential oils should be bottled in dark-colored glass, to prevent the penetration of ultraviolet light (sun) or harsh overhead room light that contains the UV spectrum, potentially spoiling the contents. Some purveyors of essential oils, particularly some who sell them at fairs and festivals, package their oils in clear glass bottles that may allow the oil to be compromised before the consumer even has a chance to use it.

24. What is a "therapeutic grade" or a "food grade" essential oil?

Labeling an essential oil "therapeutic grade" is a marketing tactic, not an official designation. Companies invent terms like "certified therapeutic grade," without specifying any agency that certifies essential oils—because no such agency or certification exists.

The only source for classification and standards for essential oils is the International Organization for Standardization (ISO), a global overseer that determines what can be called an essential oil and what is simply an oil or an extract. This organization sets rules for labeling containers of essential oils; how they should be packaged; how testing should be carried out on the oils to determine their acid value, phenol, water, and benzene content; how to determine their chromatography; and many other arcane things about them. ISO also sets very specific rules for individual oils including aniseed, bergamot, bitter orange, caraway, citronella, clementine, cypress, eucalyptus, ginger, lemon, mandarin, matricaria (chamomile), oregano, palmarosa, parsley, petitgrain, rose, sandalwood, spike lavender, sweet orange, tarragon, tea tree, vetiver, wintergreen, and others. The standards are extensive and address many of the essential oils currently on the market, but they do not define any kind of guidelines for "therapeutic grade" oils.

One of the most aggressive campaigns to make certification for therapeutic grade seem real comes from doTerra, the MLM company that states its commitment to purity on its website. The company created its own labeling to promote its Certified Pure Therapeutic Grade® (CPTG) protocol, touting its commitment to "set the standard for purity in the essential oil industry." This involves "a rigorous examination of every batch of oil, along with third-party testing to guarantee transparency." The company describes a process in which it tests each oil's chemical composition right after distillation, before the oil actually leaves its originator. It then retests the oils as they arrive at doTerra's facilities, to be sure that the oil tested at its point of origin has not been adulterated or otherwise changed between its point of origin and arrival at the bottling plant. The website goes on to detail a third round of testing, this time to ensure the purity of the oil after it has been bottled. doTerra lists eight kinds of testing that take place within its three rounds of examination: organoleptic (human sensory), microbial, gas chromatography, mass spectrometry, Fourier-transform infrared spectroscopy, chirality, isotopic analysis, and heavy metal testing. If these tests truly take place on every batch of essential oil, it may be that the company can stake their reputation on the purity of their oils—but while all of this testing certainly can help determine purity, it has nothing to do with therapeutic value. Whether this oil has any kind of therapeutic characteristics requires an entirely different kind of testing, the kind carried out on pharmaceuticals and vaccines. No essential oil marketing company conducts testing to see if these oils can be effective as treatment for so much as a scratch, let alone any illness or disease.

Other essential oils companies also use the term "therapeutic grade" to describe their products, but most of them provide no proof to back up this claim. Until science catches up with the popularity of essential oils, we cannot know if any of them have a rightful place in the modern apothecary.

"Food grade" is another marketing term, as one bottle of bergamot essential oil is as safe to ingest (in miniscule amounts) as another one. The term comes from the equipment side of food manufacturing, signifying that a machine that processes a food product can safely come into contact with that food without somehow contaminating it. Essential oil bottlers picked up this term fairly recently, to lead consumers to believe that their products can be consumed as flavorings in food or as therapeutics when added to water.

Many essential oils are used by the food industry as flavorings, because they provide a natural, sugar- and chemical-free flavor, and they are Generally Regarded as Safe (GRAS) by the U.S. Food and Drug Administration (FDA). Some essential oils companies including LorAnn Oils sell their oils almost exclusively to food manufacturers and packagers, and they make their oils available to consumers as well for cooking. While these oils are no different from the ones you may purchase for use in diffusers, the fact that they are from the same bottler that supplies them to cookie, candy, and ice-cream companies provides some assurance that they are probably fine to use in your home baking, if you follow the supplier's explicit directions for doing so. You can tell if such an oil has been approved for this use by its packaging, which must have a Nutrition Facts label that provides information about calories, additives, and vitamin and mineral content. If the essential oil does not have a Nutrition Facts label, it has not been approved for consumption.

25. What is a multilevel marketing company, and should I buy essential oils from one?

MLM companies are known by several names, including referral marketing, network marketing, and pyramid selling companies. In a nutshell, an MLM company's workforce consists of representatives who work on commission, buy their own samples and supplies, and sell products or services through networking with their potential customers in their own community, without the benefit of a brick-and-mortar store. In most of these companies, the reps are also encouraged to recruit other people to be representatives, with the goal of taking a small amount of commission for their sales. The more people the rep can recruit, the more money they make—or, at least, that's the goal.

This is not the same as a pyramid *scheme*, because the MLM actually sells a product or service, and the customer actually receives that product or service. In a pyramid scheme, the product is usually fictitious or involves an investment for a promised reward that never materializes. The representative simply takes the customer's money and tries to recruit more people to do the same. Pyramid schemes are illegal, while MLM companies are a legitimate way of doing business.

In the essential oil market, MLM reps may set up business in a beauty salon or spa and sell their products to clients of the establishment, offering a convenient way for fans of essential oils to purchase them while enjoying other services. This is all perfectly legal, and enterprising reps may manage to make a respectable income in this manner. Others purchase a booth at festivals or fairs and sell their wares to passers-by. More often, however, these salespeople rely on their friends and families to buy their products, depleting their available circle of contacts fairly rapidly.

People who become representatives of MLM companies are often expected to lay out quite a bit of money up front to buy a sales kit and enough products to sell at house parties and other venues. They may be lured to the MLM with the promise of considerable income—there's always a rep at the top of the pyramid to tell about their tremendous success—as well as a flexible schedule and the ability to "make your own hours." These companies often attract women raising children, who need an additional source of income for their families, as well as full-time students, people looking for something other than a full-time job because of family commitments, and others who are already on shaky financial ground. This, however, is how the MLM company makes most of its money, selling more products to reps than they can push to their friends and family. Returns of these sales kits may be forbidden, or the company may be willing to take them back at a fraction of what the rep paid for them. (Situation comedies love to spoof these companies—most recently including *Schitt's Creek* and *Bob's Burgers*—but the comedy comes from the truth that it can be virtually impossible to make money as a rep for an MLM.)

To be fair, there is nothing wrong with buying essential oils from an MLM company. Be aware, however, that you probably will be subjected to an aggressive sales pitch, and you may have to spend more time than you intended listening to claims about the miraculous curative properties of the oils. MLM companies often offer "starter kits" that lure you into spending much more than you planned, and you may be offered "specials," pricing with time limits designed to pressure you into a purchase quickly. If all you want is a bottle of lavender oil, you may be better served by skipping the house party where the rep will push you to spend $85 on a kit you don't need and buying a single bottle for $12 at your local supermarket or drug store.

MLM companies are sometimes the sources of outlandish claims about essential oils' ability to prevent or even cure diseases. The highest profile of these cases involved Young Living Essential Oils, which in 2014 received a warning letter from the FDA because the company promoted its products in a way that made them drugs under the Federal Food, Drug, and Cosmetic Act:

> Your consultants promote many of your Young Living Essential Oil Products for conditions such as, but not limited to, viral infections (including ebola), Parkinson's disease, autism, diabetes, hypertension, cancer, insomnia, heart disease, post-traumatic stress disorder (PTSD), dementia, and multiple sclerosis, that are not amenable to self-diagnosis and treatment by individuals who are not medical practitioners.

The letter goes on with a long list of quotes from Young Living's various websites that claim that its Thieves product, a blend of several oils, cures Ebola and a number of other viruses. Many other such statements referenced other oils and their ability to fight and cure specific illnesses.

Young Living scrubbed its website and no longer makes such claims, but many other essential oils companies do—including some claiming that their products are miracle cures for COVID-19. None of these claims have any basis in fact.

26. How do I know if the essential oil I'm buying is 100 percent pure?

The purity of essential oils comes into question fairly regularly. There is no hard-and-fast definition of "purity" in the industry; the generally accepted rule is that 100 percent of the oil must come from the plant of origin. This means that there should be no other adulterants in the bottle—that is, chemicals, water, or oils from other plants (unless the bottle contains a blend).

The top essential oils companies define their brand by the purity of their oils, with the understanding that if they added chemicals to their oils and consumers discovered this, it could lead to the end of their business. Some of these companies provide long descriptions of their testing processes on their websites as evidence that they take their commitment to purity seriously. At least one company—Rocky Mountain Oils—actually provides gas chromatography-mass spectrometry (GC-MS) test results for each of its oils on its website.

Gas chromatography separates different kinds of molecules in an essential oil to find any that are different, sending all these downstream to the mass spectrometer. The spectrometer then captures the molecules, ionizes them, and breaks them down into bits to make it possible to identify substances that should and should not be in the oil. Ionization makes these particles of matter detectable by a device called an electron multiplier. This process must be carried out in a science laboratory by trained professionals, and most essential oils companies who test their own oils before bottling engage with a third-party lab for greater impartiality and integrity.

In the end, it comes down to trusting a particular brand and its statements about the integrity of their essential oils. No governing body supervises the purity of essential oils, so any company that sells them can claim that they are 100 percent pure without having to define what this means, prove their purity, or even test the oils to be sure that the claim is true.

One of the ways purveyors of essential oils can make the case for their products' purity is through *chain of supply* (or supply chain authentication), a tracking process that requires everyone who handles the oils—from the farmers who sow the seeds and harvest the plants to the processors who distill or press the oils from those plants, and on to the bottling operation—to follow a set of strict rules, to ensure that the oils are not adulterated. If there are gaps in the process—if the oils are diverted to a place that is not part of the established chain, for example—the chain of supply is broken, and the oils would have to undergo rigorous testing to be sure that they have not been contaminated or changed in some way.

Companies that maintain an unbroken chain of supply often advertise this on their websites to emphasize their commitment to purity and quality. Young Living, for example, calls the process its Seed to Seal program, requiring trusted suppliers to follow specific protocols for growing and handling the oils it sells. doTerra's description of its CPTG purity testing process (see question 24: What is a "therapeutic grade" or a "food grade" essential oil?) also fits the definition of a chain of supply, with the added step of rigorous testing at each step.

Adulteration of oils is a common practice for some companies that offer oils at budget prices. Most consumers are not likely to send their oils to laboratories for GC-MS testing to see if they contain chemicals other than the plant essence, so these companies often get away with selling these oils as "pure" when they are not. If you are considering buying discounted oils from an online seller, keep in mind that what you are buying may contain synthetic chemicals that dilute the potency of the oil while stretching a batch to fill more bottles.

Additionally, improper storage of essential oils can damage them, affecting their strength, scent, and value to the consumer. These oils should be packaged in dark amber glass bottles, so an oil in a clear bottle may suffer damage caused by ultraviolet rays, especially if the bottles are stored in the path of direct sunlight. Companies that sell oils at a discount sometimes bottle them in clear glass vials, a sure sign that they may not be worth whatever money you may save by purchasing them. Look for dark glass stored in a cool, dry place, well shielded from sunlight or broad-spectrum lighting that may include UV rays.

27. How do I know if an essential oil is organic?

Just about all essential oils companies claim that their oils are organic, so sorting out which are truly organic and which are suspect can be a tricky business for consumers.

The good news is that there are official rules for organic farming, and products must be certified as organic to be able to make the claim. The rules vary depending on the country of origin, however, and some countries have no regulations for organic farming. So essential oils derived from plants grown outside of the United States may not all be held to the same standards for organic certification.

In general, essential oils labeled as organic should be grown without the use of synthetic chemicals, so any fertilizer, plant food, or pesticides used in their cultivation must be from natural sources. The plants must not be grown from irradiated or genetically modified seed—that is, seed on which X-rays, UV rays, ionization, or gamma rays have been used to encourage the genes in the seeds to combine in new ways, a process used in cultivation of new plant varieties. Organic farmers are required to keep their land free of chemicals for a number of years before planting (the number varies with the country), and they need to keep records in writing of their yields, their care and treatment of the crops, and to whom they sell the harvest.

In bottling the oils, the U.S. Department of Agriculture (USDA) prohibits the use of artificial preservatives, colors, or flavors for the bottlers to maintain the organic certification. Essential oils generally do not contain these ingredients because they would adulterate the oils, negating the claims of 100 percent purity.

Most essential oil bottlers in the United States label their products as "certified organic," either using the USDA Organic label familiar to shoppers in American supermarkets or using this terminology in their own labeling. U.S. products that call themselves organic must comply with USDA regulations, whether or not they have been grown within the country, so even essential oils harvested in countries with no organic farming regulations have to follow USDA rules before an American company can label them as organic.

If a company claims that its product is organically grown when it is not, the USDA has the responsibility to contact the company and inspect the product to confirm its organic content and how much of the product actually comes from organic agriculture. Products do not have to be 100 percent organically grown to qualify for the USDA Organic label, so the USDA requires documentation to determine how much of the product's contents are organic. The agency sets three levels of organic contents:

- Products made from 100 percent organic ingredients can call themselves "100% Organic."
- Products that have at least 95 percent organic content can be labeled "organic," provided that the nonorganic ingredients have been specified as exemptions by the USDA.
- Products made from at least 70 percent organic ingredients can be labeled "Made with organic ingredients." These products are not permitted to bear the USDA Organic seal, separating them from products that have more purely organic content.

If an essential oil has less than 70 percent organic content, it cannot be called organic by its American bottler. Violation of these regulations carries hefty fines, so companies generally do not attempt to claim that their products are organic if they are not.

If buying organic products is a priority for you and the essential oils you want to buy are not labeled as organic, check the company's website to see if they are or if they at least contain some organic content. If there's no mention of it, you can rest assured that their oils are not organic. Plenty of essential oil marketers do indeed sell organic products, so you should have no difficulty finding one that meets your requirements.

Keep in mind, however, that not every oil in the company's inventory may be organic. While many companies make the claim that offering organic products is a high priority, some oils may come from plants in regions that have no organic certification of their own and may not be grown organically. It should be clear which products are organic and which are not on the company's website, but you may need to look carefully to be sure.

28. What are essential oil blends, and are they more effective than single oils?

Essential oil blends are combinations of oils selected by their companies of origin to accomplish a specific goal for the customer. These are meant to save customers time and money, bringing together as many as a dozen different oils in one half-ounce bottle—a feat that customers could only accomplish on their own by purchasing all those oils separately and mixing them together themselves.

Let's take an example: Revive Essential Oils' highly regarded "Sleep" blend contains geranium, ho wood, coriander seed, lavender, Roman chamomile, lemongrass, lemon peel, ylang-ylang, jasmine, and rose essential oils and, as of this writing, sells for $15.00 for a 10 mL bottle. Should the customer choose to purchase all these oils separately from Revive and mix a similar blend by hand, the oils would cost a staggering $166.50, making the blend a more economical buy if getting a better night's sleep is the ultimate goal. Granted, they could mix this blend many times over by purchasing all the oils, but the blend becomes the safer bet to see if it works before spending all that money.

Commercially available blends also have the advantage of being mixed by experts in a laboratory, so they have been balanced properly to maximize the scent and avoid clashes between incompatible fragrances. This is not necessarily true of essential oil blend recipes that you may find in books or online, written by users without the extensive experience or resources of olfactory professionals.

If you would like to mix your own oils to create blends, it helps to have an understanding of which oils mix well together and which are likely to overpower one another or, worse yet, just smell bad when combined.

Essential oils fall into eight general categories (also referred to as families): citrus, floral, herbal, camphor, mint, spice, resin, and wood. Blends usually contain a cross section of oils from three of these categories, so the resulting scent is not overpoweringly from one category or another. A blend of orange (citrus), nutmeg (spice), and pine (wood), for example, might provide a surprisingly fresh and uplifting combination when diffused, while a blend of jasmine (floral), lemon (citrus), and myrrh (resin) may have the ability to create a relaxed, comfortable sensibility. The choices are endless and entirely up to the user, based on their own favorite fragrances among the many available.

The question of whether blends are more effective than single essential oils has intrigued some scientists in recent years, leading to a handful of studies on their effects on various viruses and other pathogens in laboratory settings. One such study (Wu et al., 2010) found that a doTerra essential oil blend called On Guard, containing wild orange peel, clove bud, cinnamon leaf and bark, eucalyptus, and rosemary essential oils, "significantly attenuated" the influenza virus PR8 in a test tube. A randomized, double-blind clinical trial (Ou et al., 2012) involving 48 patients with menstrual cramps tested an essential oil blend of lavender, clary sage, and marjoram in a massage cream for its effectiveness in relieving the cramps and found that the cream with oils provided more relief than a cream that contained no essential oils.

Two studies involving animals showed some promise for the potential power of essential oil blends. In a study (Upadhyay et al., 2019), 800 one-day-old broiler chickens became the subjects of an experiment. They were fed a commercial feed product with a blend of eucalyptus and peppermint essential oils in it, to prevent intestinal lesions, weight loss, and other side effects of a vaccine for coccidiosis, a parasitic infection common in chickens. The chicks that ate the feed with the essential oil blend showed an increase in body weight and fewer intestinal lesions, especially when the feed also contained supplemental vitamin D3. This may have little impact on human use of blends containing peppermint and eucalyptus, but it's good news for chicken ranchers supplying broilers to supermarkets.

In the second study, researchers looking for an alternative to antibiotics for livestock tested 90 weaned piglets on one of three diets: a control diet without antibiotics, a diet with antibiotics, and a diet with a blend of essential oils. They found that piglets that received essential oils grew at similar rates to the piglets that received antibiotics, and both had less diarrhea than the pigs in the control group. They concluded that supplementing the pigs' diet with essential oils could be just as effective as antibiotics, because it increased the antioxidative activity in the pigs' digestive system. This does not mean that essential oils can be substituted for antibiotics in humans, however, as we can draw no conclusions about the oils' ability to fight and kill bacteria in the human body from this study, as this was not its objective.

Much more research will be required before we can make a definitive statement that essential oil blends are more effective than single oils in any medical situation. What we do know is that a diffusion of an essential oil blend provides pleasant scents that may have the ability to improve mood and encourage relaxation.

29. Why are some essential oils very expensive, while others are lower cost?

The cost of various essential oils to the consumer can be confusing, especially when one company's lavender essential oil is twice the price of another's. There are several reasons for this paradox.

First, each plant species has a number of different varieties, all of which may be called by the same common name, but that may come from an entirely different plant. Most lavender essential oil may come from the plant *Lavandula angustifolia*, for example, while some may come from the cultivar *Lavandula hybrida*, a less common plant that the company bottling it may feel justified in selling at a higher price point. This does not make one product better than another; it's simply the choice of that company to sell a different variety.

Staying with lavender for a moment, shoppers find that bottlers may purchase their lavender essential oil from a range of sources, both domestic and overseas. Oils purchased from Bulgarian farms may be less expensive than the ones that come from France, while lavender farms in the United States may offer their own oil and pass on the cost of growing, harvesting, and distilling the oil to the customer in higher pricing.

It's easy to scan across a Google screen's ads and think that all the bottles of lavender essential oil displayed there are the same size, but companies may offer bottles ranging from 0.5 ounce to 2 ounces or higher, with higher prices for the larger bottles. Some bottlers use milliliters rather than ounces, offering bottles at 10 mL, 15 mL, 60 mL, and higher. Each increment changes the price point, of course, which may not be immediately apparent until shoppers actually click on the image and see the size of the bottle.

We have compared lavender to lavender, but some essential oils are a good deal more expensive than others because they are harder to grow and harvest. As common as the name sounds, rose essential oil is one of the most expensive in the world, because the plant yields so little of its essence in the distillation process. According to Luxatic, a website that follows luxury living, acquiring rose oil requires four tons of rose petals for every pound of oil extracted. Indeed, bottles of rose essential oil may provide as little as 2.5 mL for a staggering $70—and Revive, which offers this barely affordable vial, notes that it took 52.5 pounds of rose petals to provide these few drops of scent.

In some cases, an oil can only be obtained in one region or country, giving the grower or exporter the upper hand in setting high pricing for the entire world. Sandalwood, for example, is a popular scent from a tree that grows only in India, making an ounce of it a $500 purchase. (Young Living gets its sandalwood from the Kona Sandalwood Reforestation Project in Hawaii, and just 5 mL [0.17 ounce] of it costs customers more than $130.)

Several essential oils are much in demand by the fragrance industry and are highly prized for their exotic and rare scents. Tuberose absolute, frangipani, and champaca absolute will never appear in the local drugstore, and they are part of the reason for the high prices of top-shelf perfumes. A single ounce of champaca essential oil is well out of reach of the middle-class paycheck, running the purchaser more than $2,200 and making it the most expensive essential oil in the world.

Despite the wide range of prices for various oils, consumers may find companies that offer all their oils at very similar prices. Shoppers should see this lack of disparity between oil prices as a clanging alarm bell to shop elsewhere, as these standardized prices ("any oil for $8!") almost certainly signal that these bottles do not contain pure essential oil. It is simply not possible for a bottler to purchase lavender, rose, and neroli essential oil— another in the world's most expensive, extracted from the blossoms of the bitter orange tree—and offer them all for the same low price unless they have been diluted significantly with alcohol or even with water.

As with any purchase, it pays to be an informed shopper when purchasing essential oils. A little homework will reveal that pricing between the largest companies is not very different, and discount pricing from bulk marketplaces may not actually be a bargain. Look for botanical names and descriptions of origins (written to be enticing and poetic but also to provide facts about where and how the oils were obtained) to feel confident that the oil you decide to purchase is actually what the purveyor says it is.

30. How is the marketing of essential oils changing, and what will this mean for consumers?

With the rise in popularity of essential oils in the 2010s, the oils have made the transition from products available only through house parties or specialty retailers into the mainstream. Essential oils can now be purchased individually in supermarkets and in drug, discount, and big box stores, making them readily accessible to anyone who wants to try them.

This means that consumers have the opportunity to purchase essential oils without the pressure of a salesperson hovering at their elbow, encouraging them to buy a pricey "starter set" that contains many more products than they will actually use. A customer who simply wants to spritz some lavender essential oil in the bedroom to improve sleep can buy a bottle of lavender without feeling an obligation to buy other scents at the same time. Even the least adventurous consumer can dabble in essential oils without putting too serious a dent in the weekly budget.

This also means that more consumers will have questions about essential oils, their various uses, and the methods for using them correctly. Finding reliable information can be tricky (hence the purpose of this book), with a great deal of misinformation rising to the top of Google searches, and even more filtered through friends, neighbors, and family who may know just enough about essential oil usage to be dangerous to others. For a product line so often linked to therapeutic and medicinal use, essential oils generally do not come with dosage or usage instructions, leaving a great deal of the responsibility for administering them safely to the consumers themselves.

In addition, essential oils now occur in hundreds of food products, especially those labeled "natural" or "organic." Many essential oils in foods are used for their fungicidal or antimicrobial properties, not to flavor the foods, and they are applied in carefully measured amounts to do the job required. For careful consumers who read ingredient lists on the packaged products they buy, the appearance of essential oils in these lists may send the message that all essential oils are safe for consumption, a piece of misinformation that has the potential for harm.

Essential oils are especially prevalent in cosmetics, including perfumes, skin care lotions, shampoos, conditioners, and styling products. They lend pleasant scents to these formulations, but they also can provide nourishment for hair and skin, making products that contain them particularly desirable. Marketing messages about "natural essential oils" in these products lure consumers into believing that "[i]f it's natural, it must be good." While this may be true for most of these products, consumers need to be cautious about buying them specifically because they contain essential oils, as the oils the manufacturer selected may not actually do what the company says they do.

With essential oil sales expected to grow at a rate of as much as 9.65 percent by 2025, according to *Fortune Business Insights*, this is a time in which consumers need to be more informed than ever about essential oils and their use in packaged products, as well as their capabilities as over-the-counter therapeutics. Consumers are likely to encounter essential oils and their accessories—diffusers, sprayers, storage bottles for blends, and so on—in many aspects of their lives, from hair salons and spas to endcaps at the grocery store checkout counter. Those who want to experiment with the oils will need real information about what they are, what they can do, and what they cannot achieve for the user.

As in all purchases, consumers need to be aware of what they are buying and why, especially when buying individual essential oils or commercial blends. These oils are not cheap (at least, not when they are the real thing), so consumers need to do some homework before spending money on an oil that does not do the thing they most want it to do.

Home Use

31. What are carrier oils, and why should I use them?

Carrier oils are oils that do not turn to vapor when exposed to oxygen. These oils are used to dilute the essential oils for topical use, as essential oils can be irritating to the skin because they are so highly concentrated.

Most carrier oils are obtained by pressing the fruit, seed, or nut of a plant and are found in supermarkets, as they are used primarily in cooking. Some are more popular for use as carriers of essential oils because of their own fruity or woody/nutty scent, which combines well with the fragrances of many essential oils. Almond, grapeseed, coconut, jojoba, apricot kernel, avocado, sunflower, rosehip, argan, and even olive oil are all popular choices to "carry" the essential oils of your choice to your skin.

It is important to get the proportions right when mixing essential oil with carrier oil. Starting with 2 tablespoons of the carrier oil, add just 15 drops of essential oil, and store this solution in a tightly covered, dark glass bottle (essential oils can eat right through plastic, so be sure to use glass). If this does not provide enough of the essential oil's fragrance for you, add just five drops at a time and mix completely until you achieve the level of scent you seek.

Before using this liberally on your skin, perform a simple skin "patch" test to be sure that you are not sensitive to this particular combination. Carrier oils rarely cause an allergic reaction unless the user is sensitive to tree nuts, but an essential oil may well do so, even when added to a carrier. To perform the test, place a few drops of the oil solution on the inside of your elbow or wrist. Cover this with a Band-Aid, and leave it for 24 hours. When you remove the bandage, you can see if the area has developed redness or any kind of rash. If the area is clear, the oil solution is likely safe for use on your skin; if your skin is irritated, do not use the oil. (If you have a tree nut allergy, do not use a carrier oil pressed from nuts—almond, argan, walnut, apricot kernel, and other nut oils all may aggravate this allergy.)

Some carrier oils are more effective than others for skin care, and some have properties that make them preferable for hair care as well. Oils with high fatty acid content, including coconut, avocado, and olive, provide nutrients that assist in skin care. Argan and rosehip oils bring vitamin A to the skin, which has been found to help restore sun-damaged skin and stave off the aging process. Argan, apricot kernel, and grapeseed oil also supply vitamin E, which has been found effective in some studies for helping to heal scars and burns. Some carrier oils are known for their ability to be absorbed quickly into the skin—jojoba, apricot kernel, sweet almond, olive, argan, grapeseed, and sunflower all have this property, making them good choices for dry skin care and overall healthy skin.

Depending on the essential oils you plan to mix with your carrier oil, the additional scent that the carrier oil brings to the mix may make one carrier better than another for your purposes. Coconut, jojoba, almond, apricot kernel, olive, argan, and avocado oils have strong fragrances of their own, as does rosehip—though it does not smell like rose essential oil, supplying an earthy, slightly nutty smell instead. Most of these carrier oils bring a woody, nutty aroma to the mix, making them a good pairing with earthier scents; coconut, of course, blends well with floral and fruity essential oils for a tropical combination. By contrast, grapeseed and sunflower oils have almost no scent of their own, so they do not interfere with the combined fragrances of the essential oils you add to them.

32. What are the different ways that essential oils can be used at home, and are certain methods better than others?

Adding essential oils to household cleaning products can provide clean, fresh scents as well as a boost in grease-cutting power. Lemon has long been a popular addition to commercial products (think Lemon Pledge), so it should come as no surprise that lemon essential oil tops the list of useful alternatives to furniture polishes made of synthetic chemicals.

A favorite formula for people who choose natural products over packaged ones involves mixing a few ounces of jojoba oil in a glass or metal spray bottle with drops of lemon, sandalwood, and lemongrass essential oils (use twice as much lemon as the other two). This makes a useful polish that will shine up wooden tables and chairs.

For cleaning a wood floor, any of the evergreen oils (pine, Douglas fir, juniper, etc.) mixed with a few tablespoons of unscented liquid soap can replace more caustic products. Add a few drops of lemon essential oil as well to lift greasy spots, and pour the whole mixture into a bucket of hot water as you would a powdered cleanser or a squirt of liquid cleaner.

Essential oils can be used to make your own laundry detergent, fabric softener, air freshener, bathroom cleanser, or window cleaner, eliminating the need for racks of specialized household products. Find recipes for any of these products online or in books including one of mine, *Essential Oils and Aromatherapy: An Introductory Guide.*

Making your own skin and hair care products allows you to take advantage of essential oils as well. Rosehip, carrot, lavender, neroli, evening primrose, and myrrh essential oils, for example, are well known for their benefits to the skin and are often found in high-end, commercially available products. Tea tree essential oil (also known as melaleuca) has been proven in several peer-reviewed studies to be as effective as the pharmaceutical benzoyl peroxide in treating acne (see question 40: How are essential oils used in skin care products?).

Lavender, rosemary, and a wide range of oils from flowering plants can provide hair conditioner with an added boost, while birch or sage essential oils added to white vinegar and vegetable glycerin can help correct oily hair.

Of course, essential oils can be diffused in the home as air fresheners, deodorizers, disinfectants, and to lift mood, promote relaxation, and clear the mind. Sachets made with essential oils can be tucked into drawers and closets to provide a pleasant scent to clothing or linens (see question 39: How are essential oils used in scented products like air fresheners?).

Mixing your own products for home use can be fairly simple, requiring not much more than white vinegar, baking soda, unscented soap, glycerin, and a few carrier oils. It is important to remember that essential oils will dissolve some plastics, so all products made at home should be stored in dark glass or metal bottles and should be kept in a cool, dry place to prevent spoilage or evaporation. (Some recipes call for the products to be stored in the refrigerator.) Try making a small amount of the product and using it a few times to be sure that it achieves the purpose for which it is intended, before investing in a lot of an expensive oil and mixing up a big batch.

33. Can essential oils be taken internally?

Ingestion of essential oils is a controversial subject, hotly debated between purveyors of essential oils, longtime practitioners of various therapies that use them, the global medical community, and governmental organizations.

There is no specific process for preparing essential oils to be used in food or taken as if they were medicine. Essential oils that marketing companies call "food grade" are no different from the essential oils offered for aromatherapy—nothing has been done in their harvest, distillation, or bottling to somehow make them safer to use in food.

The U.S. Food and Drug Administration (FDA) does provide guidance for the use of specific essential oils as flavorings in packaged foods. A few essential oil suppliers—LorAnn Oils in particular—work directly with the packaged food industry, providing essential oils that have received a rating from the FDA of Generally Regarded as Safe (GRAS) for use in food manufacturing. These companies may call their oils "food grade," but this simply refers to their GRAS rating, not to any special or different process to prepare them for use in food.

Many users believe that because essential oils are natural and organic, they are safe for internal use. This logic, however, breaks down when we consider all the things in nature that are not only natural and organic but also harmful and even deadly: many kinds of mushrooms, poison ivy, poison oak, water hemlock, nightshade, white snakeroot, oleander, and tobacco, for starters. While some may scoff at the comparison of lemon or orange essential oils to these hazardous plants, the fact is that science has not delved deeply into the relative safety of most essential oils, especially those sold by companies that do not otherwise produce any kinds of food products.

LorAnn Oils, a seller that supplies "super-strength flavor" essential oils to the food production industry, has been certified by the Safe Quality Food (SQF) Institute, a division of the Food Industry Association, to show that it has made a "commitment to a culture of food safety and operational excellence in food safety management." Companies that receive SQF's certification must follow the specific code detailed in an extensive rulebook, to prove that they "meet and exceed all industry, customer, and regulatory requirements so they can remain competitive across sectors." Consumers can purchase flavorings made with essential oils from this company, but even LorAnn cautions that its oils "are safe when used sparingly . . . but should not be ingested undiluted."

What if you add a few drops of an essential oil to water and drink it, as many practitioners suggest you should? Despite all the recipes you may find online, and all the books that suggest this as a common and well-established practice for remedying a long list of ailments, the short answer is this: no, you should not ingest any essential oil in this manner, nor should you rub a drop of oil on your gums, drip it into an ear, or insert a swab containing an essential oil into any other bodily orifice. At the very least, some essential oils are irritating to mucous membranes and will cause the equivalent of chemical burns. At worst, some oils can cause a toxic reaction, with symptoms ranging from anaphylaxis (throat swelling, gagging, choking, shortness of breath) to nausea, vomiting, and diarrhea. The severity of the reaction may vary from one person to the next, making your own decision to ingest an essential oil something of a gamble.

As stated earlier in this book, essential oils are not regulated by any governing body, so they have not undergone the extensive third-party testing that any food product must endure before it goes on the market. The companies that market essential oils for aromatherapy, use in skin creams, and as additions to household products are not required to pass through any review of their products to make them safe as food. No matter what the multilevel marketing representative tells you, adding these oils to your own food or drink may put your own health in jeopardy.

34. What is the best way to store essential oils, and can they go bad or lose their potency?

Essential oils should be packaged and stored in dark glass bottles—usually amber—to protect them from ultraviolet (UV) rays in sunlight or full-spectrum lighting fixtures. UV rays can compromise the oils by encouraging oxidation, a process through which the oil loses its fresh color and some of its fragrance. This may happen more quickly if the bottles do not achieve an airtight seal when closed. Some oils, including citrus oils (lemon, lime, orange, tangerine, etc.), may become cloudy when they have oxidized, a sure sign that it is time for a new bottle.

A study (Treibs, 1960) discovered that when essential oils are stored in metal containers, impurities in the metal can be released into the essential oil. Copper and ferrous iron, in particular, promote oxidation, and later studies (Choe and Min, 2006) have found that storage in metal will initiate the oxidation process. This, coupled with the fact that many essential oils eat through plastic, makes the glass bottle the best option for storage.

If you mix your own blends, lotions, or cleaning products using essential oils, be sure to store them in airtight containers. As essential oils are volatile—meaning that they turn to vapor when exposed to oxygen—leaving a lotion scented with an essential oil in an open container will allow the lotion to lose its scent fairly quickly.

Storing bottled oils inside a box or cabinet in a cool place will protect essential oils from both light and heat, two factors that can change the chemical composition of the oils. While some suppliers sell lovely display stands for essential oils to be kept on a counter or dressing table, leaving the oils where they can be exposed to sunlight will not extend their lives. Keeping them in a carrying case in a vehicle also shortens their usefulness, as a car's interior can rise above 100 degrees as it stands outside on a sunny day. In fact, essential oils can oxidize in a matter of days with increasing heat, spoiling your entire apothecary quickly.

It can be fairly easy to tell if the oils have lost their strength or if they have oxidized significantly. Some look faded, and opening the bottle does not bring the burst of fragrance we expect, even when sniffed. Some oils become skin sensitizing as they oxidize, developing new characteristics that can cause redness and rashes on skin. Prolonged exposure to light and/or heat can even cause some oils to morph from liquid into a solid, a state from which the oil will not recover.

Some practitioners recommend that essential oils be stored in the refrigerator. This certainly will prevent oxidation, though chilled oils will require some time on the counter to warm up before use.

Like all substances in the world, even if stored properly, essential oils will eventually degrade on their own. This happens at different rates for different kinds of oils. Research has shown that some essential oil blends deteriorate more slowly than single oils, whether mixed by the seller or the consumer. This may be because the oxidation of one oil may affect the rate of oxidation of others, either speeding it or slowing it down (antioxidating) depending on the oils in the blend. Studies of this phenomenon date back as far as 1948, but more research is required to fully understand what may be happening to keep blends from degrading as quickly as individual oils.

35. What precautions should people take when using essential oils?

In addition to the potential for some essential oils to be toxic, or to cause allergic reactions on skin or when taken internally, some oils pose specific hazards that consumers who are new to essential oils may not expect.

Some plants carry molecules that make them photodynamic when applied to skin, releasing free radicals that can react to ultraviolet light and cause skin damage. These oils are more reactive than others to UV light, so they can cause the skin to which they are applied to burn when exposed to the sun. These are known as photosensitizing oils, and they should not be applied to skin that will be in direct sunlight within 12 hours of application. They should not be blended into sunscreen, as they will have the opposite effect, and may cause the most severe sunburn you've ever had. These oils include angelica, bergamot, bitter orange, cumin, dill, ginger, grapefruit, lemon, lemon verbena, lime, orange, tagetes, tangerine, and yuzu. (Note that when bergamot, bitter orange, grapefruit, lemon, lime, orange, and tangerine essential oils are obtained from the plant through steam distillation, they are not photosynthesizing—only the ones obtained through cold-pressing have this hazardous effect. Check your supplier's website to see how the oils in your collection have been obtained.)

While it may seem that aromatherapy and candles create a nice atmosphere for relaxation, be sure to keep your essential oils away from open flame. Essential oils are volatile, so they can catch fire in an instant. Don't store them near a gas stove or a fireplace, either.

Some essential oils encourage relaxation to a point of sleepiness, so using them in a diffuser in your car while driving could be a hazardous choice. These include benzoin, carnation, chamomile, geranium, hops, hyacinth, lavender, linden blossom, mace, marjoram, neroli, nutmeg, ormenis flower, petitgrain, sandalwood, spikenard, valerian, vetiver, and ylang-ylang, but this may not be a complete list. You may find that another essential oil puts you to sleep in minutes when you diffuse it in your bedroom. If this is the case, be careful not to use this oil in your vehicle, especially while driving.

Perhaps most important, if you are using an essential oil as therapy or as a curative for an illness, do not use it in place of a medication prescribed by your doctor. While a number of essential oils may be diffused or applied topically as a complementary therapy to modern medicine, no essential oil has been proved by science to be a cure for any disease or illness. Using an oil instead of a prescription may prolong the illness or even allow it to get worse while you wait for the essential oil to work. Do not endanger your own health or the health of a loved one by waiting and waiting for an oil to make magic—even if you believe that God sent us the oils and that we should trust in natural remedies. God also sent us brilliant scientists and doctors to help protect our health and cure illnesses; trust in this goodness as well as the fruit of the vine to pull you through.

36. How long does it take for an essential oil to work?

The effectiveness of any essential oil depends on which oil is in use, the way it is being used, the purpose the user expects it to serve, and the method of its application. The answer to this question presupposes that the essential oil will accomplish whatever task to which it has been set; in many cases, it may not "work" at all.

Oils used in a diffuser to fill a room with scent, promote relaxation, or lift a mood may do so in minutes, as this is one of the simplest missions they fulfill. While there is not a lot of scientific research on aromatherapy, studies have shown that fragrances stimulate receptors found in the nose, which in turn send signals through the nervous system that reach the limbic system. The limbic system triggers emotions, so it creates a positive response to the pleasant scent. This response can happen in seconds. If you are waiting for a diffused essential oil to help you sleep, it will probably take longer, but studies suggest that scents can help relax the mind and body and encourage sleep in 15 or 20 minutes. (If your diffusion does not seem to be effective, consider trying a different oil or a combination of oils.)

For topical applications, the effects of essential oils combined with a carrier oil, lotion, cream, salve, or other products depend on the regularity with which they are applied. Just as a facial moisturizer takes days or weeks to correct dry, oily, or combination skin, a moisturizer that contains essential oil will take just as long to work. Users may experience some immediately soothing heat, coolness, or softening generated by the oil, depending on which essential oil(s) have been added to the lotion, but a more complete effect usually will take longer, and the carrier substance will have the greater effect on the skin. The exception is tea tree oil, the only essential oil proven by science to be as effective on teenage acne as some over-the-counter treatments; it can clear an outbreak of pimples in a few days when applied directly to the affected skin.

Essential oils used in household products like all-purpose cleaners, detergents, and soaps provide a clean, fresh scent on contact, usually from the citrus, floral, or evergreen families. As these products begin cleaning immediately, you can expect quick results—especially if you are using lemon essential oil, which has long been recognized as a cleaning agent.

If you are using one or more essential oils to remedy a medical issue, it can be difficult to predict when the oils will work or even if they will work at all. (See questions 13 through 22 for scientific research findings about essential oils and illnesses.)

Some oils have a fairly rapid effect when used to treat symptoms of an illness or injury, such as sinus congestion caused by the common cold. Inhaling steam containing a few drops of eucalyptus essential oil can help open up the upper respiratory tract, for example, providing temporary relief. Lavender essential oil placed on the temples or used in a diffuser has been found to be an effective treatment for some headaches, particularly those caused by tension, and may provide relief in half an hour or so. Tea tree essential oil quiets the pain of a sunburn on contact, especially when mixed with an aloe vera gel. These are well-known remedies for ailments that, while uncomfortable, are not life threatening, and users often welcome the opportunity to use a nonpharmaceutical option to achieve some relief from symptoms.

There is no research that reached the conclusion that essential oils are effective against any potentially fatal disease or illness, however, so it is currently impossible to say that any essential oil will work on a cancerous tumor or that it will prevent or correct heart disease, stroke, cancer, lung disease, kidney disease, liver disease, or any other such condition. It may be that using essential oils as a complementary treatment may assist with symptoms like pain, anxiety, and even nausea, but only a medical professional can determine if such a usage would be appropriate in your or your loved one's specific situation. Some oils even exacerbate life-threatening conditions like liver disease or estrogen-dependent cancers, so they are contraindicated in those cases.

For any situation in which you choose to use essential oils as a complement or alternative to pharmaceuticals, set a time limit. If you do not see the results you seek in 24 hours, go back to follow your doctor's instructions. Without scientific research that proves that the oil you've employed will treat or cure your condition, there is no way to know how long it will take for the oil to "work" or even if it will have any effect. Do not endanger your own life or the life of another by pursuing healing in a blind alley.

❖

Essential Oils in Everyday Products

37. How are essential oils used in food?

With the recent trend toward "clean" eating—choosing food products that contain no chemical additives like artificial colors, flavors, thickeners, or preservatives—essential oils have found significantly increased popularity among food manufacturers and packagers. Not only do consumers perceive essential oils to be natural, but most of the major bottlers and marketers of the oils in the United States take the step of certifying them as organic by the U.S. Department of Agriculture (USDA), making them appropriate for use in products labeled as "natural" or organic.

Essential oils can be found in the ingredient lists for many candies, chewing gums, ice creams, baked goods, pudding, gelatins, chocolates, frostings, marinades, salad dressings, herbal teas, and even soups and other canned products. These oils are not packaged in the same way as the ones consumers buy for aromatherapy: they have nutritional information on their bottles, a requirement for any ingredient to be used in the food processing industry. They also may contain vegetable glycerin or propylene glycol, organic compounds that make them easier to blend into foods on the scale of a manufacturing and packaging facility. This does not make the oils any less organic or natural—it simply adapts them to the needs of the industry. The FDA supplies food manufacturers with instructions for the use of each oil, including how much of the oil is safe as a food additive and which oils can be used in this manner.

Essential oils packaged for use in food in consumer households are available from selected essential oil bottlers (LorAnn Oils is the best known of these), and they come with specific instructions for their use in food, to be sure that consumers do not overuse the oil and end up with unpleasant results.

Recently, the food manufacturing industry has made significant investments in research to determine if any of the claims about essential oils' antimicrobial properties could be true. The results have led some food packagers to begin replacing synthetic preservatives with specific essential oils that have shown promise. A list of oils grown and harvested in Italy and tested by Pellegrini et al. (2018) turned up some "interesting biological potentiality," suggesting that they may be candidates for "natural biopreservatives," with coriander, fennel, garlic, oregano, peppermint, rosemary, salvia, savory, and thyme in the lead for this. A study (Ballester-Costa et al., 2017) found that of the many varieties of thyme essential oil on the market, two species—*Thymus zygis* and *Thymus capitatus*—demonstrated antibacterial activity when tested on beef. Hsouna et al. (2017) found that citrus lemon essential oil showed antioxidant activity against *Listeria monocytogenes* bacteria in beef and actually inhibited the bacteria's development.

A study by Houda Banani et al. (2018) discovered that apples treated with thyme or savory essential oils showed resistance to a gray mold, *Botrytis cinerea*, a common nuisance in the apple industry. Banani concluded that the oils sparked the apple's own defenses against the mold, making it produce a gene to protect itself. Additional research on thyme and savory oils by Santoro et al. (2018) found that treating peaches and nectarines with a vapor containing these two oils could control brown rot, improving the storage quality of the fruit.

A study (Nowotarska et al., 2017), published in the journal *Foods*, examined the antimicrobial abilities of cinnamon and oregano essential oils against *Mycobacterium avium paratuberculosis*, a pathogen that infects both livestock and humans and is found in milk, cheese, and meat. The study determined that the oils targeted the cell membrane of the pathogen to begin the destruction of the bacteria, providing insight into how these essential oils' antimicrobial properties work. This could be a useful step in determining if these and other oils can be useful as food preservatives.

So why not just throw out all the chemical preservatives and use essential oils instead? The transition is not so simple as that, because food production uses heat, light, and air—three things that can degrade essential oils rapidly and rob them of their effectiveness. The industry continues to explore ways to make use of the oils' antimicrobial properties, including actually adding the oils to the food packaging rather than to the recipes. Creating edible films and coatings that protect the food with their antioxidant and antimicrobial properties may be one solution, especially if such a thing takes nothing away from the flavor or appearance of the food itself. Research is ongoing, with new discoveries and innovations expected.

38. How are essential oils used in cosmetics?

Beauty products and perfumes often contain essential oils, usually for their fragrances, as they have the ability to mask the unpleasant odors of fatty acids, oils, and surfactants required in the manufacturing of cosmetics. Any product used for cleaning, perfuming, changing the appearance, protecting, or correcting the odor of the human body that is not specifically labeled "fragrance-free" or "unscented" may contain essential oils.

Just as these oils have been deployed in food products recently because of their potential antimicrobial properties, the cosmetics industry has also embraced them as substitutes for chemicals. Using one or more essential oils in a cream, lotion, shampoo, conditioner, or gel may negate the need for other preservatives, antioxidants, and antifungal agents, allowing cosmetics manufacturers to tout their products as "all natural." In addition, a number of essential oils provide desirable benefits to consumers.

Helichrysum essential oil, for example, stimulates blood circulation in the skin and reduces the appearance of wrinkles, making it highly prized in skin creams and lotions. Lavender essential oil's well-known ability to calm inflammation and heal minor abrasions helps it find its way into many different products. Dental product manufacturers often include German chamomile essential oil in mouthwashes and toothpastes, because of its ability to reduce inflammation in the gums and mouth; it goes without saying that a number of mint oils including peppermint, spearmint, and wintergreen are important to the dental industry as well.

Neroli (bitter orange) essential oil, one of the most expensive in the entire essential oil industry, is much in demand in the perfume and soap industries, both for its fragrance and for its ability to refresh aging skin. Rose essential oil is nearly as expensive and equally desired for perfumes, body lotions, creams, ointments, and soaps, earning it the nickname "liquid gold" in the industry. Much less expensive but valued for its ability to calm dandruff and nourish hair, rosemary essential oil provides its scent to all kinds of bath products as well as shampoos and conditioners.

While cosmetics are not to be taken internally, their topical application can pose a risk of allergic reaction, particularly the malady known as contact dermatitis—an itchy, red, blistery patch where the allergen was applied. This sometimes occurs when a substance applied to the skin is photosensitizing, a well-known trait of a number of essential oils (see question 35: What precautions should people take when using essential oils?).

The cosmetics industry has gone to great lengths to be sure that the essential oils used in its products do not cause this or other allergic reactions, and the packaging of most cosmetics suggests that users perform a patch skin test before using them liberally on the body. It is entirely possible, however, that consumers will find that they are sensitive to a specific ingredient in a new cosmetic preparation, so the patch test can help users avoid the discomfort of contact dermatitis.

A 2017 review of cosmetic products and literature by Sarkic and Stappen at the University of Vienna discovered that many cosmetic products do not contain essential oils at all, but rather counterfeited versions used for their similar scent and lower price point. These copies were adulterated with the addition of a single raw material, using "nature-identical" compounds isolated from other oils or adding synthetic compounds devised in a laboratory. Some used less expensive oils from the same plant family or oils taken from a different part of the plant that may not carry the same compounds that give the original oil its benefits. Consumers have no way to know if the essential oil advertised on the product's packaging is the real thing, unless they take the product to a laboratory and have gas chromatography-mass spectrometry (GC-MS) testing performed on it—an expensive process at best, unless the consumer happens to be a lab technician. This creates a level of safety for the cosmetics manufacturer in making the claim that the product contains pure essential oil.

It is unfortunate but true that consumers who buy cosmetic products have no way to know if the product they selected actually contains essential oil. The only clue may be the price: the cheaper the product, the lower the likelihood that the scent comes from essential oil and not from a synthetically diluted counterfeit.

39. How are essential oils used in scented products like air fresheners?

There is a short answer to this: most air fresheners and other room-scenting products do not contain essential oils. Commercially manufactured, scented products including aerosol sprays, plug-ins, and solid air fresheners usually contain a range of synthetic chemicals, combined to create the nebulous substance defined simply as "fragrance" on the product label.

Why don't we all know that these products are full of unnatural substances? The manufacturers are not required by law to disclose the ingredients on their labels. Faced with considerable criticism and bad press for the number of preservatives and chemicals these products send into the air, some manufacturers have been brave enough to publish the lists of individual chemicals used in these products on their websites. The list for Glade Radiant Berries solid air freshener, for example, reads like this: "2,6-dimethyl-7-octen-2-ol; 2-t-butylcyclohexyl acetate; benzyl acetate; dipropylene glycol; ethyl 2-methyl-1,3-dioxolane-2-acetate; ethyl 2-methylbutyrate; ethyl methylphenylglycidate*; hexamethylindanopyran*; hexyl cinnamal*; ionone; isobutyl methyl tetrahydropyranol; limonene*; linalool*; methyl 2-ethylhexanoate; methyldihydrojasmonate; raspberry ketone."

Warnings from various state departments of health throughout the United States note that the laboratory-produced ingredients found in air fresheners—chemical fragrances, propellants, solvents, deodorizers, and fragrance distribution mechanisms known as phthalates—are all irritating at best, and some can even be carcinogens.

Scented candles are an alternative to air fresheners, but these also pose challenges. Most of these candles are molded using paraffin wax, which is made from petroleum and can produce chemicals including toluene, benzene, and soot when lit. These can irritate the respiratory systems of people with asthma, chronic obstructive pulmonary disease, and other lung issues. At worst, they can cause cancer in heavy users. Most of these candles also contain synthetically concocted fragrances like the ones in air fresheners.

It is possible to buy air fresheners that do not contain all these substances, though they are few and far between. A few candle and air freshener makers—most notably Ananda Natural Soy Candle Company, Big Dipper Wax Works, Grow Fragrance, and Rare Earth Naturals, among others—create their products from all plant-based materials and use essential oils to provide the scents. These products are more expensive than the ones you may find in the supermarket or discount store, but this is because essential oils and other natural ingredients are often more costly than their synthetic counterparts.

If you are experimenting with essential oil fragrances that you like, you may be better served by mixing your own air fresheners or even making your own candles. Adding a few drops of your favorite essential oil to three ounces of water creates a "spritz" that can be kept in a glass or metal spray bottle and sprayed in whatever room you like. Using an essential oil diffuser will be just as effective. You may find that the scent of an essential oil lingers longer than any synthetic air freshener.

40. How are essential oils used in skin care products?

Most commercially made skin care creams, lotions, ointments, and toners do not contain essential oils. Selected products from a few brands—Kiehl's, Annmarie, and Honest Beauty among them—do contain essential oils (specifically lavender or rose), while a number of other organic brands use the less costly leaf extracts from various plants instead of the parabens, phthalates, and other chemical compounds that appear in most mass-produced skin products.

Next to aromatherapy, skin care is one of the most popular uses of essential oils, and there is considerable evidence that specific oils have a positive effect on various kinds of skin. They must not be applied directly to skin without significant dilution in an unscented lotion or cream or in a carrier oil such as coconut, apricot kernel, or avocado oil.

Some essential oils can actually harm skin, even when mixed in a lotion or carrier oil. Citrus oils are photosensitizing, which means that they are activated by sunlight and can cause damaging sunburn. Avoid applying lemon, lime, orange, tangerine, bergamot, grapefruit, or mandarin essential oils on your skin in any form. Mint oils including balm, pennyroyal, peppermint, spearmint, and wintergreen essential oils can be irritating to skin, creating additional inflammation.

Making your own skin cream or lotion is not as tricky as you may think. Blend a quarter of a cup of jojoba oil, two tablespoons of coconut oil, two tablespoons of beeswax, and a tablespoon of shea butter to create the base, and heat this in a one-quart saucepan over very low heat, or use a double boiler. When all the ingredients have melted together, add a few drops of the essential oil you have chosen for your skin type (more on this in a moment), and mix well. Pour this into a glass or metal jar, and store in a cool, dark place. It is advisable to try a skin test before you apply this to your face: rub a little of the lotion into the skin on the inside of your elbow, and wait 24 hours to see if your skin reacts. If you do not see redness or a rash develop, you can begin using your homemade lotion on your face and body. (You can also purchase an organic, unscented lotion from a natural foods store and add the essential oils you have selected.)

People with dry skin may add a few drops of lavender, chamomile, or sandalwood essential oils to this base to add some benefits while providing the pleasure of a naturally scented lotion. Palmarosa, rosewood, and carrot seed essential oils are also credited with a positive impact on dry skin.

If oily skin is your issue, clary sage, rosemary, geranium, neroli, and frankincense essential oils help balance the skin's alkaline levels. Research has shown that rosemary is particularly useful for reducing inflammation, such as the redness that can come with sunburn, contact dermatitis, or rosacea.

For people with sensitive skin or those with eczema or other chronic skin maladies, lavender and sandalwood are particularly soothing. Avoid oils that are very acidic, like lemon and other citrus oils as well as lemongrass.

Tea tree oil has been proven repeatedly through scientific research to be very effective in combatting acne. Other essential oils have the ability to reduce the inflammation and swelling of whiteheads and "spots": lemon and lemongrass essential oils are naturally astringent, making them good substitutes for prescribed preparations like salicylic acid. Cinnamon essential oil's antioxidant properties can help shrink pimples and relieve redness as well.

Many commercial skin care products boast of their ability to reduce the outward signs of aging. A few essential oils may provide some of this protection: sandalwood's anti-inflammatory properties have been noted in early studies, while clary sage, a proven antioxidant, may protect the skin from damage by free radicals. A study (Zarfeshany et al., 2014) determined that pomegranate essential oil (as well as the fruit's flesh and juice) can help reduce oxidative stress and combat free radicals, key components in the aging process. Lavender, the all-purpose essential oil, also provides antioxidants to the skin. A 2015 study revealed that ylang-ylang essential oil, coveted for its floral scent, actually can help rebuild proteins and fats in the skin to reduce the appearance of wrinkles, even as it conquered free radicals. Rosemary, rose, and frankincense all have positive effects on aging skin as well. Guidelines for use of these oils on skin include mixing three to six drops of the essential oil per ounce of a chosen carrier oil—coconut, almond, avocado, and apricot kernel are all good choices for use on skin.

Most essential oils marketing companies suggest consulting a dermatologist before you use any essential oil on your skin, especially if you have a skin condition such as eczema. It is always a good idea to speak with a medical professional before starting a new therapy—especially essential oils, as they are not regulated by the FDA or any other entity. Physicians who specialize in the science of skin care will have knowledge of any complications patients have had with using essential oils.

41. How are essential oils used in hair care products?

Oils have had a place in hair care for centuries, especially for people living in areas where water was scarce. The enriching properties of oils restored body and shine to dry, damaged hair, and the addition of sweet-smelling essences from flowering plants provided pleasant scents as well as additional conditioning.

Today, shampoos, conditioners, styling products, and even massage creams to prevent hair loss all may contain essential oils. Consumers should read the ingredients on any hair product carefully, as the ones that list many botanicals may be derived from plant matter rather than essential oil. (There is nothing inherently wrong with this; the plant's leaves, bark, flowers, or seeds may provide the same general properties that the essential oil would provide. The potency of these properties may be stronger, however, if the product contains essential oil.)

A number of essential oils have been shown to improve hair growth, encourage the development of new follicles, reduce dandruff, increase hair's strength, and make hair look and feel healthier overall. A study published in *Toxicological Research* in April 2016 (Lee et al., 2016), for example, determined that lavender essential oil had "a marked hair growth-promoting effect" comparable to minoxidil, the prescription medication promoted for hair growth. In 2014, in the same publication, a Korean team (Oh, Park, and Kim, 2014) determined that peppermint essential oil "could be used for a practical agent for hair growth," as it showed the most hair growth after four weeks—more, in fact, than minoxidil. Blending several drops of these oils with a teaspoon of a carrier oil (jojoba or coconut oil is popular for hair care applications) and massaging this into the scalp can promote hair growth; practitioners suggest leaving it on the scalp for five minutes before rinsing.

Cedarwood, ylang-ylang, clary sage, and lemongrass essential oils have the ability to balance the scalp's production of oils, helping to control hair loss, dandruff, and generally oily hair. Geranium essential oil is often used as well, as it strengthens hair and helps keep it from breaking and splitting. The easiest way to use these is to pour a bit of shampoo into the palm of the hand and add a drop of each of these oils into the shampoo. Massage this into your hair and scalp, and wait five minutes before rinsing.

Tea tree essential oil, one of the all-purpose oils that becomes central to an essential oil afficionado's collection, can be found in some commercially manufactured anti-dandruff shampoos and conditioners. Any product can be used to combat dandruff by adding up to 10 drops of tea tree essential oil to a 12-ounce bottle of shampoo. Shake well, and use this daily.

As a conditioner to help hair look and feel healthy and strong, several essential oils are well known to add shine, moisture, and nutrients to hair. Chamomile and lavender essential oils promote shine and manageability, while sandalwood helps to mend dry, split ends. Adding a few drops of these oils to a bottle of conditioner can make their application simple; make your own conditioner using Moroccan argan oil or jojoba oil as a base and adding the essential oils. Wrap your head in a towel or put it up in a shower cap for 10 minutes to allow the oils to do their job.

About every essential oils company has its own recipe for a deep conditioner, but in general, they recommend a basic mix of one tablespoon olive or jojoba oil to three tablespoons coconut oil and eight total drops of a variety of oils that are known to be good for hair: lavender, peppermint, cedarwood, clary sage, rosemary, ylang-ylang, geranium, and chamomile. Mix this well, work it through your dry hair, wrap your hair in a towel, and let it work for 15 minutes. Rinse, shampoo, and style as usual. This weekly treatment provides all sorts of hair-stimulating benefits, from hair growth to shine and body.

42. How are essential oils used in medicine?

In the United States, only a few over-the-counter (OTC) medical products feature any form of essential oil. As the oils are not considered pharmaceuticals and their marketers cannot make any health claims about them without running afoul of the FDA, American OTC drug manufacturers do not use them in their products.

This is not the case in other countries, however. Essential oils play a role in Ayurveda medicine, a practice that originated in India and is believed to be as much as 40,000 years old. Texts that date back 6,000 years discuss the use of oils derived from plants to maintain the *dosha*, the three functional energies in nature that govern the human body: *kapha* (earth and water, including structure, growth, and cohesive functioning); *pitta* (fire, including metabolism, becoming, and transformation); and *vata* (air and space, for movement, communication, and separation). Ayurveda employs aromatherapy to balance each of the *doshas*, using scents that are opposites of the qualities of each *dosha* to provide grounding and restore equilibrium. This holistic approach to healing makes the Ayurveda apothecary very different from the many medications we find in our own bathroom cabinets. The Ayurveda Awareness Center in Applecross, Western Australia, provides this list:

> The Ayurvedic pharmacy is filled with aromatic plants that are well known throughout the world, as well as its own collection of unique species. Herbaceous species include tulsi (holy basil), coriander, sages, fennel and mints. Aromatic roots include vetiver, valerian, and calamus. Flowers include roses, jasmine, champa, marigolds and lotuses. Tree species include sandalwood, cedar, agarwood, pine, and eucalyptus. Many resins are utilized, including frankincense and

guggul, a species of myrrh. Ayurveda is also rich in spices, including cinnamon, cardamom, black pepper, long pepper, ginger, nutmeg, and clove. Several aromatic grasses are found, such as lemongrass.

The description ends with the note, "If you have a known medical condition, consult your physician before using this."

Traditional Chinese medicine also incorporates essential oils into its methodology. This practice uses aromatherapy and aromatic herbs to reg-ulate the *wei qi*, or the body's defense against disease—in other words, the immune system. Essential oils come into play to treat the *shen*, the emotional health of the patient (some translate this as the spirit). Practi-tioners of Chinese medicine believe that aromatherapy takes advantage of the link between the sense of smell and neurological systems throughout the body—specifically, the centers of memory and emotion and even cir-cadian rhythms. Treatment with diffused essential oils might be employed for insomnia, anxiety, irritability, difficulty concentrating, and some forms of more serious mental illness like depression. "On a practical level, we find that scent is one of the most fast-acting, nontoxic, and effective tools to both elevate the mood and calm the mind," said David Crow, L.Ac., founder of Floracopeia Aromatic Treasures and practitioner of traditional Chinese medicine, in an interview with Ben Zappin of Five Flavors Herbs. Crow noted that aromatherapy can aid in rest and improve sleep cycles, as well as improving alertness and concentration.

Several European countries stock essential oils alongside manufactured medications and remedies in pharmacies, with France taking a leadership role in this. Some essential oils in France are packaged in gel capsule form to take orally for a wide range of illnesses and conditions, though this practice is becoming less prevalent there. France and Germany, where essential oils have been available to consumers to use medicinally for quite some time, are both showing movement toward more conservative prescriptions, with a greater emphasis on the results of scientific studies and peer-reviewed research to determine the most effective applications. Ingestion is falling out of fashion; instead, aromatherapy and topical application are the more recommended methods for use among those who practice alternative medicine.

43. What new uses for essential oils are on the horizon?

The field with the greatest growth potential for essential oils is in food packaging, where essential oils may provide a natural, organic defense against mold, fungus, and bacteria that cause food to spoil. Thousands of studies published in scientific journals since the early 2000s have tested the claims made by essential oil growers and manufacturers that these oils have antimicrobial properties. Many have returned positive results in vitro—that is, in laboratories in which an essential oil's components are combined with a specific bacterium or fungus in a test tube. Further studies are likely to involve the application of essential oils to actual food products, to see if the oil (or its components) actually preserves the food's freshness as well as inorganic preservatives can.

While the food industry funds this research, it has become clear that some essential oils may have actual medicinal advantages as well. It may be that we will see essential oils in OTC medicinal products such as anti-bacterial creams and ointments or that they will find uses in hospital settings to help guard against "hospital illnesses," infections like staph and MRSA that patients often pick up from roommates or their environment during a prolonged hospital stay.

In 2019, orange essential oil turned out to be the oil with the greatest market share, purchased in massive quantities by the cosmetics industry. Recent studies have shown that orange oil has the ability to act as a skin toner, fading stretch marks and quieting dermatitis and acne. This property has helped the oil find its way into many skin care products, providing its fresh, familiar scent as well as its therapeutic value—and its value as a beauty product has grown as well, as young women choose products with orange essential oil for their nails and hair as well as their skin.

Studies have also shown that peppermint essential oil has legitimate value in relieving pain, because of the high concentrations of menthol it contains. This penetrating vapor makes the oil a viable product or additive in the personal health care industry, for treatment of muscle pain, headaches, and cold and sinus congestion.

Discoveries like these may be on the horizon for many essential oils, now that interest in research is receiving so much support and funding from major industries. The more this science reveals about the oils, the more prevalent they will become across a broad spectrum of products.

Overall, however, the greatest growth market for essential oils remains their potential for home use. As we learn more about the hazards caused by the many synthetic preservatives, fragrances, and aerosols available in a wide range of household products, many people have turned to products that do not contain so many manufactured chemicals. Cleaning products that replace synthetic scents with lemon or pine essential oil, for example, may draw a greater audience of people who are concerned about their own family's health as well as the health of the environment and the planet.

In personal care, essential oils have moved beyond scenting perfumes and hand lotions and now can be found in toothpastes, mouthwashes, soaps, shampoos, and baby products. Men's grooming products have become a growth market for the oils, as they were more often used in women's products to provide floral, herbal, and fruit scents; the expansion of the market for body sprays with earthy and spicy scents has been a boon to essential oils and now represents a new opportunity for further growth.

Potential Hazards and Side Effects

44. Are some essential oils toxic?

Some essential oils are highly poisonous and should not be taken internally for any reason. Most of these are not available to consumers in the United States, because they are not on the Generally Regarded as Safe (GRAS) list. Some of these are carcinogenic, and others cause instant chemical burns when used on skin. If you have an old collection of oils that contain one or two of these, or if you see any of these advertised online, do not purchase or use them:

- Ajowan
- Balsam of Peru
- Bitter almond
- Cade oil crude (prickly juniper)
- Calamus (sweet flag)
- Camphor
- Colophony
- Costus root (kuth)
- Fig leaf absolute
- Horseradish
- Jaborandi
- Massoia bark
- Melaleuca bracteate

- Melilotus
- Mustard
- Ocotea
- Parsley seed
- Rue
- Santolina
- Sassafras
- Savin
- Southernwood
- Styrax gum (oriental sweet gum)
- Tansy
- Tea absolute
- Thuja
- Tonka bean
- Verbena
- Wormseed
- Wormwood

Other common essential oils, including wintergreen, nutmeg, eucalyptus, and tea tree, can cause serious complications if swallowed. They are safe for topical use but must not be ingested.

Some essential oils are potentially hazardous for people who have heart problems, especially high blood pressure. Peppermint essential oil has been found to react with calcium channel blockers (medications like amlodipine) and make them more potent, potentially dropping a patient's blood pressure to a dangerously low level. At the same time, ingestion of peppermint essential oil can cause heart palpitations and increased heart rate, which can be equally dangerous for a cardiac patient. Research suggests that these patients should avoid hyssop, rosemary, sage, and thyme essential oils as well, as these can stimulate the heart.

People who have epilepsy should avoid essential oils that may increase the potential for seizures: camphor, fennel, hyssop, rosemary, sage, and spike lavender (*L. latifolia*, not the much more common *L. angustifolia*, which is normal lavender). These oils have a convulsive effect. Likewise, health professionals and some essential oil sellers recommend that people with cancer refrain from using aniseed, basil, bay, clove, cinnamon, fennel, ho leaf, laurel, nutmeg, and star anise essential oils, especially if the cancer is estrogen dependent, as these oils can stimulate this hormone.

Some essential oils can be toxic to liver function if swallowed. While we have already reviewed the case against ingesting any essential oil mentioned previously, these have a documented ability to do harm to the liver: aniseed, basil, bay, buchu, cassia, cinnamon, clove, fennel, and tarragon. Conversely, there is no documentation that prohibits their use in aromatherapy or for topical application.

45. How can essential oils affect hormones?

The effects of essential oils on the body's hormones can be either positive or negative, depending on the user's age and gender.

Ramsey et al. (2018) completed a study at the National Institute of Environmental Health Sciences and presented its findings at the Endocrine Society's annual meeting that year. Ramsey determined that lavender and tea tree oil both contain constituents that can disrupt the endocrine system—the glands that produce hormones that control metabolism, sexuality, reproduction, sleep, growth, and other bodily functions. The study found evidence that these two essential oils can behave like estrogen and that can suppress testosterone, both of which can affect children during puberty. An increasing number of cases have found male gynecomastia in boys—abnormal growth of breasts—when the boys use lavender and/or tea tree oil topically. When they stopped using the oils, their breasts returned to normal. Ramsey and his colleagues broke down lavender and tea tree oils and chose eight components for further study, testing them on human cancer cells in test tubes and finding that some of them "demonstrated varying estrogenic and/or antiandrogenic properties . . . consistent with endogenous, or bodily, hormonal conditions that stimulate gynecomastia in prepubescent boys," as a March 2018 news release from the Endocrine Society reported.

The constituents tested by Ramsey's team occur in "at least 65 other essential oils," the report continued, suggesting that pubescent boys should avoid essential oils or risk the possibility of endocrine system issues.

This study met with a great deal of pushback from entities including the Australian Tea Tree Industry Association, which considered the Ramsey study "sensationalism" that caused "undue concern for consumers . . . around the world." The association pointed to other, older studies that "demonstrated the flaws in linking TTO . . . to endocrine disruptor activity." *Endocrine News* reported in July 2018 that Ramsey and his team "acknowledge that what was presented at [the conference] and the subsequent publication isn't the final word on the health effects of essential oils. Their work warrants further investigation, and there are more questions to answer."

On the other end of the age spectrum, essential oils are often considered a partial remedy for the symptoms of menopause. A study published in the September 2008 issue of *Evidence-Based Complementary Alternative Medicine* (Hur, Yang, and Lee, 2008) used aromatherapy massage with lavender, rose geranium, rose, and jasmine essential oils on 25 menopausal women, with a control group of 27 women who did not receive such massage. The experimental group reported "significantly lower total menopausal index" than the control group, with reductions in symptoms of menopause including hot flashes, night sweats, melancholia, achiness, muscle pain, and stiffness. "However, it could not be verified whether the positive effects were from the aromatherapy, the massage or both," the researchers noted.

Practitioners of aromatherapy usually recommend a range of essential oils to relieve menopausal symptoms: clary sage, fennel, lavender, jasmine, and especially geranium for depression; juniper and rose oils for pain relief; angelica, lavender, and rose for anxiety relief; and neroli to aid sleep. A systemic review published in the *Journal of Menopausal Medicine* (Khadivzadeh et al., 2018) sought a remedy for the sexual dysfunction symptoms of menopause—particularly sexual desire—and found that "it is possible to improve the standardized mean difference (SMD) of the sexual desire up to 0.56 in the aromatherapy group compared with the control group," using aromatherapy with a combination of lavender, fennel, geranium, and rose essential oils. However, the study concluded, "The findings of the present review should be presented cautiously because of the corresponding limitations such as the lack of a standardized tool, the lack of intention-to-treat reporting, the low study amount, and the short-term follow-up."

Likewise, a study published in the journal *Complementary Therapies in Clinical Practice* (Heydari et al., 2018) conducted a double-blind clinical trial on 62 university students to determine the effects of *Citrus aurantium* (bitter orange blossom) essential oil on premenstrual syndrome (PMS). The students filled out a questionnaire about their premenstrual symptoms and then inhaled either the fragrance of this essential oil or an odorless sweet almond oil during the onset of PMS. The students who had aromatherapy with the essential oil saw an improvement in their symptoms, while the control group did not.

Essential oils also may have both positive and negative effects on pregnant women, which we will explore in question 48: What should pregnant women know about using essential oils?

46. Can essential oils be used with children?

The answer to this question depends on the child, the child's health, and the medications the child might be taking for specific health conditions.

Some parents swear by essential oils and their ability to help a child sleep, reduce a child's anxiety or pain, ease a sour stomach, or open clogged sinuses for a child who suffers from allergies or frequent colds.

The idea of using a natural substance rather than a manufactured one can seem very attractive to a parent, especially one who emphasizes organic foods, natural fibers, and other healthful alternatives.

Not every child tolerates the use of these oils in the same way, however, and some find them more irritating than beneficial. If this is the case with your child, discontinue use of the essential oil immediately.

With some precautions, some essential oils may be used safely with children:

- **Always dilute the oils.** Oils should never be used directly out of the bottle on children's skin, as they can injure the skin. Add a drop or two of the oil to a carrier oil or lotion before applying. Experts at Johns Hopkins Health System in Washington, DC, recommend a 0.5 to 2.5 percent dilution for any essential oil to be used on a child.
- **Do not add essential oils to bath water.** The oil will not dissolve into the water but will remain concentrated on the water's surface. This can irritate a child's skin.
- **Children should not take any essential oil internally.** Do not add the oil to a child's food or beverage, or give the oil as a medicine. The oils are too concentrated for a child's consumption.
- **Avoid peppermint oil with children.** Peppermint oil increases the risk of seizures in children under 30 months old.
- **Use very sparingly.** Children may be more sensitive to scented products than adults are, so these products can trigger breathing issues, rashes, and other allergy-like reactions. For example, using a scented lotion and then turning on a diffuser that contains essential oils can be harmful to a child.
- **If your child will be in the sun, do not apply essential oils to the skin beforehand.** Citrus oils and others are photosensitive and can cause the skin to burn if the child plays in direct sunlight.
- **Don't use any essential oil near the child's eyes, nose, or ears.** Some publications and websites suggest using essential oils as a remedy for a child's ear infection or a cold, but applying an essential oil to a child's mucous membranes can cause adverse reactions in some children.
- **Choose your oils wisely.** Buy oils from a reputable source rather than shopping for the cheapest or most easily accessible brand. Some products that call themselves essential oils actually contain adulterating chemicals, including alcohol (see question 26: How do I know if the essential oil I'm buying is 100 percent pure?).
- **Talk to your pediatrician before using any essential oil on your child.** Your doctor has seen the results of essential oils used with children, so they can advise you about what may be safe or unsafe for your child's health.

If the child develops a rash, irritated skin, a headache, nausea, wheezing, or any other difficulty breathing, discontinue use of the essential oil and call a doctor.

Essential oils are not an alternative to medical care for any serious condition or illness a child may develop. No essential oil has been found to be a replacement for antibiotics, vaccines, or medical therapies. Always follow the advice of your child's doctor.

47. Can essential oils be used with pets?

Using essential oils in a household with pets can be hazardous to the pet's health, depending on the pet's contact with the oil or its vapor and the oils being used.

Diffusing essential oils near pets in a closed room can cause respiratory distress for the pet, especially if the diffusion continues for more than a few minutes. Keep in mind that dogs and cats have a much keener sense of smell than humans do, so even a small amount of an essential oil's scent can be overpowering to an animal. Pets also do not have the same response to scents as humans, so the essential oil that calms a person's anxiety, for example, may have exactly the opposite effect on a dog or cat.

An assortment of essential oils is actually toxic when used on a pet's skin or when a dog or cat accidentally ingests them (if a few drops spill on the floor and the pet licks them up). Do not use these around cats or dogs, as these can be very harmful or even fatal:

- Anise
- Cinnamon
- Citrus
- Clove
- Eucalyptus
- Garlic
- Juniper
- Lavender
- Oregano
- Pennyroyal
- Peppermint
- Pine
- Sweet birch
- Tea tree
- Thyme

- Wintergreen
- Yarrow
- Ylang-ylang

Research has shown that cats do not have the ability to process the phenols in essential oils—one of the parts of the oil that produce the scent. This means that when a cat inhales the diffused essential oil or swallows some of it, it can cause liver damage or liver failure, as well as respiratory failure and even seizures. Dogs can develop respiratory issues, increased stress, central nervous system issues, and changes in behavior from inhaling diffused essential oils or ingesting products that contain oils.

"If you can smell the aroma of the oil, that means that there's oil in the air and can result in respiratory distress," the Cabbagetown Pet Clinic in Toronto, Ontario, explains on its website.

Veterinarians warn pet owners to watch for signs of a pet's negative reaction to an essential oil:

- Watery nose
- Watering eyes
- Skin redness
- Lips or gums turning red
- Vomiting
- More drooling than usual
- Coughing or wheezing
- Difficulty breathing or panting
- Tremors or wobbling
- General lethargy and fatigue
- Pawing at the face or mouth
- Low heart rate
- Low body temperature

While these are known to be toxic to animals, essential oils not listed here may affect your pet negatively as well. If the cat or dog displays any of these symptoms, take them outside immediately into the fresh air; if the symptoms do not clear up quickly, act fast: call your veterinarian immediately, or take your pet to an animal medical center's emergency room. If your pet has rolled in the essential oil or has some on its paws, wash it off right away to keep the pet from licking it and ingesting enough to be harmful. If you start to see symptoms of poisoning, get your pet to an emergency clinic. Bring the bottle of the essential oil you were using with you, so the vet, clinic, or medical center knows exactly what the animal has encountered. As a further precaution, keep your essential oils stored in a safe place and away from your pet, just as you would keep medications away from your children.

All of this being said, you will find that some veterinary practices recommend the use of natural products containing essential oils, confusing the issue somewhat. Even the American Kennel Club markets some products that contain essential oils on the toxic list. The fact remains, however, that natural flea and tick repellents that use essential oils may or may not be effective, and some have the potential to be harmful to your pet. If you use one of these products, check the ingredients to be sure that the product does not contain one of the oils on the list provided previously. Many recipes for homemade flea and tick repellent online suggest peppermint and tea tree oil as ingredients, but these are indeed toxic to pets and should not be used. (Flea and tick products that contain citronella and cedar, two known insect repellants, may be more effective in warding off bugs without as much potential for harm.)

If you use essential oils regularly in your home and you want to know the best thing to do to protect your pets, consult your veterinarian. The effects on various breeds may be different, so it is important to talk with a professional to be sure that you are doing what is best for your dog, cat, or whatever animals share your home with you.

48. What should pregnant women know about using essential oils?

If you are pregnant and you wish to use essential oils to relieve some of the symptoms of discomfort that may accompany pregnancy, talk with your obstetrician before doing so. Your doctor will have the most recent information of what may be helpful, harmful, or simply ineffective in helping you during this time.

Doctors agree that women in the first trimester of pregnancy should avoid the use of essential oils, to be sure that the fetus is not exposed to any substance that may turn out to be toxic. Some essential oils can cause uterine contractions during the first trimester, and some may have an adverse effect on the baby during this critical developmental period. While there is very little research on the effects of essential oils on developing fetuses, it is common sense to follow this advice from the obstetrics community.

In the second or third trimester, however, some essential oils may be useful in diffusions to calm anxiety and assist with relaxation and sleep, some may be helpful when added to a carrier oil for massage, and others may be preferable to harsh chemicals for housecleaning. Obstetricians generally warn against taking any essential oil internally, as this may have a toxic effect on the fetus and even cause a miscarriage.

Some studies indicate that lavender essential oil may actually help reduce the pain of labor and delivery when diffused in the delivery room. Mirzalinajmabadi et al. (2018) conducted a systematic review and meta-analysis of all the studies to date, published in the *Journal of Obstetrics, Gynecology & Cancer Research*. The researchers concluded that "aromatherapy with lavender reduced labor pain the active phase." The studies examined used lavender essential oil in a diffuser, a practice that can certainly be employed with ease in the delivery room. Massage with lavender oil diluted in a carrier oil also may have the desired effect.

Some other essential oils may help calm anxiety, improve relaxation, relieve nausea, and reduce the pain of headaches during pregnancy. Rose essential oil can be effective in promoting sleep as well as calming frayed nerves. Roman chamomile essential oil has been found to relieve migraine pain, according to a 2014 study; Roman chamomile and geranium oils help reduce anxiety, including during labor.

Oils considered safe during pregnancy include the following:

- Argan
- Bergamot
- Cardamom
- Cypress
- Eucalyptus
- Fennel
- Frankincense
- Geranium
- Ginger
- Grapefruit
- Lavender
- Lemon
- Lemongrass
- Lime
- Mandarin
- Myrtle
- Neroli
- Patchouli
- Pomegranate
- Roman chamomile
- Rose

- Rose otto
- Rosewood
- Sandalwood
- Sweet orange
- Tea tree
- Ylang-ylang

The list of essential oils to avoid during pregnancy includes the following:

- Aniseed
- Arnica
- Basil
- Birch
- Bitter almond
- Boldo leaf
- Broom
- Buchu
- Calamus
- Camphor
- Cassia
- Cedarwood (thuja)
- Chervil
- Cinnamon
- Clary sage
- Clove
- Coriander
- Costus
- Deertongue
- Elecampane
- Horseradish
- Hyssop
- Jaborandi leaf
- Juniper berry
- Melilotus
- Mugwort
- Mustard
- Nutmeg
- Oak mass
- Oregano
- Parsley
- Pennyroyal

- Pine
- Rosemary
- Rue
- Sage
- Sassafras
- Savin
- Tansy
- Tarragon
- Thuja
- Thyme red
- Tonka
- Wintergreen
- Wormwood

Peppermint often lands on the list of forbidden essential oils during pregnancy, but recent studies suggest that when used strictly for aromatherapy, peppermint essential oil can reduce the nausea that many pregnant women experience. A diffusion of lemon essential oil can also have this effect when used regularly.

49. Can people be allergic to essential oils?

When a person's immune system overreacts to what would normally be a harmless substance, they are having an allergic reaction. The substance may be an airborne particle like pollen, dust, or pet dander; allergens can also be found in food (peanuts, melon, soy, wheat, or milk, for example), in materials that come in contact with the skin (like a soap or detergent), or in a gaseous state, like air pollution or a diffused scent. The body's reaction can range from sneezing and coughing to stomach cramps, ringing in the ears, and hives. Some severe allergies can cause anaphylaxis, a fast-moving process that swells the throat and tongue, making it difficult or impossible to breathe.

People and pets can be allergic to essential oils. The symptoms will vary from one person to the next, and chances are that the allergy will be to one specific oil or a family of oils, not to the entire compendium of essential oils. Just as people may be allergic to ragweed but have no adverse reaction to goldenrod, a person may be allergic to lavender essential oil, for example, but not peppermint or sandalwood.

An allergic reaction to essential oils can take one or more of many forms. Some people who mix essential oils into their hand or body lotion may develop contact dermatitis, a red rash that surfaces when the essential oil comes into contact with the skin. The rash may become worse, resulting in cracked skin, oozing sores or blisters, and pain or a burning sensation. Contact dermatitis may be a sign of general irritation rather than an allergy, perhaps from using too much essential oil without blending it into a carrier oil or lotion. Stop using the oil until you have determined if you have an allergy or simply failed to mix the essential oil and carrier oil thoroughly.

Reviews of skin patch tests in 2010 and 2012 revealed that certain essential oils are more likely to cause an allergic reaction than others. It is interesting to note that some of these are among the most popular essential oils in use worldwide, which may be the reason we know more about their potential for allergic reaction than others:

- Clove
- Jasmine
- Lavender
- Lemongrass
- Peppermint
- Sandalwood
- Tea tree
- Ylang-ylang

This list is not exhaustive; other essential oils may cause allergic reactions, but they may not have been studied for their potential as allergens.

Beyond a red rash, some people develop hives, also known as urticaria, from using essential oils. It can be very difficult to pinpoint the cause of hives, as a wide range of factors can make them appear, from food allergies to stress. You can tell them from contact dermatitis fairly easily, however, as hives are raised welts on the skin, often several inches long and very itchy. They may come and go over time, making it particularly tricky to determine if they appeared as a reaction to an essential oil—and they can be caused by things like foods that have no direct contact with skin.

To help predict if an essential oil will cause an allergic reaction, perform a simple skin test. Place a drop or two of the essential oil in a teaspoon of carrier oil, and rub this on the skin on the inside of your elbow. Cover this and wait for 24 hours, and then remove the covering (gauze or a Band-Aid will do) and see if you have a rash, swelling, or hives. If you don't see any irritation, chances are the essential oil is safe for you to use. If you do have a reaction, you can wash the area with mild soap and water to remove the last of the essential oil; if the reaction is severe, try running cold water over a washcloth and applying this to the affected area to soothe the burning and itching.

Essential oils used in a diffuser can generate an allergic reaction—most likely sneezing, coughing, nasal or chest congestion, and runny nose. This can be particularly pronounced in people with severe hay fever and seasonal allergies, chronic sinusitis, asthma, or chronic obstructive pulmonary disease (COPD). If you suffer from any of these health issues, it may be wise to forgo an essential oil diffusion in your home or workplace. Try a short-duration diffusion (perhaps five minutes) in a room to see if it affects your ability to breathe or if it triggers an allergic response. If your body reacts negatively to the essential oil diffusion, stop using it.

Some essential oils can cause eye irritation, but this is more a factor of people forgetting to wash their hands after adding an essential oil to a diffuser. If you have essential oil on your hands and you touch your eye, your eye may become irritated, producing a burning sensation. Flush your eye with water if this happens; if the irritation does not clear up in a few minutes, call your ophthalmologist.

50. How much usage of an essential oil is too much?

What is "too much," when there are no instructions to tell us? We have no advice from medical professionals about dosage, so we can only guess at the amount required to achieve the desired effect. This makes it fairly easy to misuse essential oils.

Consumers turn to books like this one, word of mouth, multilevel marketing (MLM) representatives, or online sources for information. It seems likely that eventually, essential oils will come to be under the jurisdiction of a government organization—most likely the U.S. Food and Drug Administration (FDA)—as their potential for treating illness and diseases becomes better understood. Once this happens, we will receive official guidance on dosages, safe amounts to use for specific issues, and other information that will reduce the fog that shrouds information about these oils.

Online sources provide information that consumers can interpret as they choose, take as gospel, or ignore entirely in favor of some other source. We all know that the internet provides as much misinformation as fact, and many sources of this misinformation have become highly skilled at making themselves look like medical professionals or authorities. The result: essential oil users take their practice into their own hands, selecting whatever information seems to make the most sense to them—whether it comes from true experts, quacks, dabblers, or boastful for-profit MLM companies.

Some essential oils companies push the notion that if a little is good, more must be better. This is not the case for oils—in fact, their concentrated potency makes it important to use them very sparingly. As a rule of thumb, you are using an essential oil too much if any of the following things happen:

- Vapor from essential oils hangs in the air in your home, office, or car, scenting the air long after you have turned off the diffuser. If you can still smell the essential oil days later, you used too much to begin with. (A good rule to follow: run the diffuser for 20 minutes, and then turn it off until the following day.)
- You or someone in your household develops breathing issues while using a diffuser with essential oil.
- Someone in your home develops a rash or hives after washing with soap that contains an essential oil or from wearing clothing washed in a detergent that contains one or more oils.
- You try to treat an illness with essential oils instead of seeing a doctor, and the patient's condition does not improve or becomes worse.
- Anyone in your home takes an essential oil by mouth and has a reaction—stomach cramps, nausea, vomiting, and so forth.
- Anyone in your home develops seizures, hyperactivity, narcolepsy, rapid heartbeat, headaches, migraine, or asthma, whether or not these conditions can be traced back to use of essential oils.
- You or someone in your home experiences bodily changes caused by a hormonal imbalance: breast growth in boys, for example.

In extreme cases, a user will have a reaction to overuse that is so severe that it will have lasting effects for years. Once a consumer has an initial reaction to an essential oil, they will continue to have that reaction every time they come into contact with that oil.

Beware of MLM reps who tell you that a reddened area on your skin is a "detox" reaction rather than an allergic response or who insist that you should add a few drops of essential oil to your post-workout water bottle every day. These people are sales associates focused on selling as much product as they can rather than on your personal health and well-being. Instead of taking their word for the best ways to use an essential oil, do your own homework about the risks and benefits of any oil you would like to try. A licensed aromatherapist can help you choose the right oil for your specific goals—and they will have enough experience to know how much of each oil is the right amount and how much is too much. (Try to find an aromatherapist who does not actually sell essential oils as part of the practice. As with any salesperson, they will focus on the oils that will net the most profit if the money from the sale benefits them personally.)

Case Studies

1. REMEDIES FOR THE COMMON COLD

Six-year-old Jacob wakes up one morning with a scratchy throat. He tells his mother, Miriam, and she feels his forehead and finds no elevation in his temperature. "Let's wait and see," she says, deciding to send him to school and see if he develops additional symptoms before beginning any kind of treatment.

As the day goes on, Jacob starts sneezing regularly, and he begins to sniffle. When he comes home from school with a full nose and watery eyes, Miriam sees that he probably has a cold and bundles him into bed. Soon his eyes are red, and he breathes only through his mouth. Miriam concludes that he has a cold, and as a mother who does her best to use only natural products in the care of her child, she gathers her essential oils to begin to treat the illness.

Jacob has no fever, so Miriam decides to focus on bringing him some relief from his runny nose and congestion. She heats a pint of water until steam rises and then pours it into a glass bowl and adds two drops of eucalyptus essential oil. She puts the bowl on a tray and brings it into Jacob's room and positions it on his lap in the bed. As Jacob leans over the bowl, she drapes a towel over his head and shoulders so that it also covers the bowl. At his mother's direction, Jacob closes his eyes and breathes deeply, inhaling the steam and the vapors from the oils. He continues to breathe in this scented steam for three minutes. His sinuses clear somewhat.

Miriam takes away the bowl and the towel, and Jacob relaxes in bed and breathes a little easier. They repeat this ritual twice more that day and three times the following day.

Between these treatments, Jacob becomes understandably fussy and cranky as he sits in bed feeling uncomfortable. Miriam places a diffuser in his room and adds two drops of bergamot essential oil and two drops of chamomile essential oil and lets this run in the room for 15 minutes. When she checks on Jacob later in the afternoon, she finds that he is sound asleep and concludes that the essential oils did the job of calming him and helping him get some rest.

To quiet the pain of Jacob's sore throat, Miriam makes a wrap to place around his neck. She adds two drops of bergamot essential oil and two drops of rosemary essential oil to a pint of hot (not boiling) water in a glass bowl and soaks a hand towel in the water for a few seconds. When the towel becomes fully saturated, she wrings it out, wraps it in a dry towel, and places it loosely around Jacob's neck. Jacob sits in bed and watches television while the towel's warm vapors soothe his throat. When the towel cools, Miriam removes it. She repeats this process several times over the next day or two.

On the third day after Jacob came home with the cold, he feels better, and his sinus congestion and sore throat have cleared up. Jacob returns to school, and Miriam shelves her essential oils until the next time her son has a need for them.

Analysis

Miriam chooses to use essential oils for her young son's cold because she recognizes that a cold is a temporary illness that will resolve on its own eventually. Instead of reaching for over-the-counter commercial products that contain manufactured chemicals that she feels her child does not need, she relies on heat, steam, and vapors from essential oils that are proven household remedies.

Eucalyptus is one of the active ingredients in Vicks VapoRub, a product that serves many parents in helping to clear congestion from a child's bronchial tract, as well as suppressing a cough. While the Procter & Gamble product can also relieve minor muscle aches, it contains synthetic camphor, petrolatum, and turpentine oil, ingredients that are harmless to children, but that many mothers may consider to be unnecessary chemicals to rub into a child's skin.

Research has demonstrated that using bergamot essential oil can be effective in reducing stress. A 2017 study, published in the journal

Phytotherapy Research, added bergamot essential oil to the atmosphere in the waiting room of a mental health treatment center, and researchers observed that the 57 participants in the study displayed improved mental well-being. This study provided preliminary evidence that bergamot essential oil can be a useful aromatherapy in relieving stress and improving clarity and mental health. Miriam selected this oil for the diffusion in her son's room because of her more anecdotal understanding of its use as a stress reliever, but she made a potentially solid choice.

Chamomile essential oil has long received commercial recognition as a stress-relieving scent, with chamomile tea considered one of the most relaxing flavors—in fact, many tea manufacturers recommend chamomile tea before bed to improve sleep. A study published in February 2020 in the journal *Burns* discovered that patients with severe burns who received massages with aromatic oil containing chamomile essential oil slept significantly better than patients who received a placebo massage or no massage. This suggests what tea drinkers have known for some time, that the scent of chamomile may calm people and improve their sleep.

Note that Miriam never places undiluted essential oils on her son's skin or gives them to him to take internally. Neither of these practices is recommended when using essential oils with children, and even diffusions should be used sparingly until the parent is sure that they have no adverse effect on the child. She also avoids the use of lavender essential oil, perhaps because she knows that this oil can upset the balance of hormones in prepubescent boys—and while Jacob is not old enough to experience those effects, Miriam has plenty of other oils in her apothecary that can have the same relaxing benefits as lavender.

2. BUYING FROM A BARKER

While attending a street festival with many booths and a variety of carnival "barkers" selling their wares, Jeffrey stops at an attractive display of essential oils lined up across the front of a booth. The oils, arrayed in clear glass vials of uniform size, sparkle in the sunlight and sport colorful labels that depict herbs, flowers, and other parts of pretty botanicals. The label artwork helps to guide the eye to the oil's name and claims of its capabilities. Signs in the booth advertise the oils as "100% Pure" and provide a cost breakdown with special pricing for purchasing multiple vials of oils. The pricing is the same for any oils the customer buys—$8.00 for the first one, $14.00 for two, and $20 for three bottles.

Jeffrey has heard that oregano essential oil is the top oil for shrinking and removing skin tags, a pesky annoyance for people who have reached a certain age. He has not shopped online or in stores for this oil, but now that he stands in front of this beautiful display, this one piece of information comes to mind immediately. He gets the salesperson's attention and asks about the powers of oregano to get rid of his skin tags.

"Oh yes, oregano will do it," says the salesperson, grabbing a bottle from the display and handing it to Jeffrey. "Put a drop of this on each skin tag at night, and in the morning, the tags will just fall off."

Jeffrey reaches for his wallet to pay for the oil, but the salesperson points out the special pricing for multiple vials. He touts the benefits of frankincense essential oil as an aphrodisiac and rose essential oil for dandruff control, and Jeffrey pulls out $20 to pay for all three.

That night, Jeffrey uses a cotton swab and puts a drop from the vial of oregano essential oil on each of his skin tags. He goes to bed, expecting to wake up with his skin tags ready to fall off.

In the morning, however, he checks the skin tags and finds that nothing at all has taken place. Frowning, he reapplies the oregano oil after his morning shower and also adds several drops of rose essential oil to his shampoo to activate its expected dandruff-relieving properties. He continues this practice for several days, but the skin tags do not dry up, and the dandruff does not disappear. Worse, the skin around each of the tags grows red and irritated, so he stops dabbing on the ineffective oil.

By this time, with so little success from two of the three essential oils he purchased, Jeffrey doubts that the frankincense essential oil will have any effect on his partner's libido. One evening, he decides he might as well give it a try. He diffuses the essential oil in the living room after dinner. After 20 minutes, his partner looks up and sniffs the air. "Is something burning?" he asks. "It smells like burnt bark in here."

"In a good way?" asks Jeffrey.

His partner sniffs again. "I don't think so," he replies, and returns to reading on his tablet.

Jeffrey turns off the diffuser and decides to watch something on Netflix.

Analysis

Sign after sign flashed like neon to warn Jeffrey that he had been taken in by a carnie and should not buy so-called essential oils from this booth at the street fair. Jeffrey clearly has no experience with essential oils, however, so he was easily duped into buying oils that could not possibly be pure, fresh, or effective.

First, the essential oils at this booth were all in bottles of the same size, which were made of clear glass. Real essential oils invariably come in dark glass bottles to filter out the harmful ultraviolet rays of sunlight and other full-spectrum lighting. The fact that these oils were not only in clear glass but were exposed to direct sunlight—so much so that they sparkled in the sun—would be clues to people who use essential oils regularly and understand how to keep them fresh.

The size and price of the bottles provided more clues. Essential oils often come in bottles of different sizes based on the quantity and price of the oil, especially on the high end where a tiny amount of oil can be very expensive. (Rose and neroli, for example, are often sold in quantities as small as a third of an ounce to make them affordable at all.) Some oils can be sold at fairly low prices, while others are much more expensive, based on the processes involved and the quantity of oil that can be acquired from an amount of plant material. For example, it can take four tons of rose petals to extract a pound of rose essential oil, so a 2.5 mL bottle of rose essential oil may sell for a staggering $70. Other oils require this kind of an outlay as well, so whatever the street fair salesperson hawked as rose essential oil for $8.00 could not possibly have contained this oil.

Jeffrey bought his oils on a whim, doing no research on which oils did what before he allowed himself to be snowed by this barker. While he was on the right track with the oregano essential oil being a tried-and-true method for killing skin tags, the salesperson's directions were wrong: the oil needed to be added to a carrier oil to prevent the irritation Jeffrey developed by applying it directly. Even so, whatever was in the bottle either had already been spoiled by exposure to the sun or had been adulterated and did not contain full-strength oregano essential oil—or, perhaps, both things occurred. The result was wasted money and effort.

The bottle labeled "rose essential oil" most likely contained a few drops of rosewater—a distillation of rose petals and steam, requiring much less plant matter than rose essential oil does—and perhaps further diluted with alcohol or plain water. Rosewater has no particular effect on the scalp, but alcohol can be drying to the scalp and therefore can actually aggravate dandruff. The salesperson most likely noticed that his customer had a little dandruff and grabbed the closest vial to sell to him as a remedy for this. (Lavender and tea tree essential oils have proven ability to alleviate dandruff when added to a mild shampoo.)

Finally, frankincense is not an aphrodisiac. Exotic scents like jasmine, patchouli, and ylang-ylang have been suggested for sparking the libido when diffused, though there is no scientific proof of these claims. Much more popular for its potential effect on anxiety and stress, arthritis pain and stiffness, antimicrobial wound care, and asthma, frankincense's woody, spicy aroma probably creates too heavy an atmosphere for sexual desire, perhaps even having the opposite effect entirely.

Luckily for Jeffrey, he only wasted $20 on the three vials of scented liquid, any of which he can use in a diffuser or add to a bath if he chooses. He can also pour them down the drain, and no one would blame him for this.

3. SCENTED THERAPY

Rhonda, a woman in her 30s, has battled generalized anxiety disorder for much of her adult life. She suffers from debilitating symptoms including periodic inability to get a deep, satisfying breath; dizziness; fear of crowds; trembling hands in social situations; insomnia; and other common manifestations of the disorder.

Fearful that she will lose her job and her friends if she does not get these symptoms under control, and exhausted from feeling this anxious all the time, she makes an appointment with her doctor. She describes all these symptoms, and the doctor prescribes an anxiety medication she can take as needed and a selective serotonin uptake inhibitor (SSRI) to take once daily to correct the chemical imbalance that causes this disorder. The doctor also recommends alternative therapies to help her relax, including acupuncture and aromatherapy.

Rhonda begins taking the medications and feels better, but she continues to react to some situations with the same anxiety. Long habits of shielding herself from anxiety triggers continue to keep her from resuming activities that she enjoys. She decides to give aromatherapy a try. She makes an appointment with a practitioner of holistic alternative therapies, who sees her in an office with certifications in frames on the walls, noting her membership in the National Association for Holistic Aromatherapy and her certification by the American College of Healthcare Sciences. The therapist suggests that she place a diffuser in her living room and diffuse lavender and clary sage essential oils for 15 minutes each evening. The therapist also recommends a hot bath scented with geranium and bergamot essential oils, which also provide a calming effect, and she provides Rhonda with a small spray bottle of a mist scented with lavender to spritz in her bedroom to help her relax and sleep. She also recommends several different brands of essential oils that she trusts and directs Rhonda to stores and online sources that sell them. Rhonda leaves the aromatherapist's office and proceeds to one of the stores, where she purchases the oils that she needs.

That evening, Rhonda begins her diffusion therapy while she watches television. She diffuses the oils and turns the diffuser off after 15 minutes, and half an hour later, she realizes that she does indeed feel calmer and her lingering anxiety symptoms have lessened. She also notes that she feels more cheerful and optimistic, which she decides must be a pleasant side effect of the aromatherapy. Later in the evening, she fills the bathtub with hot water and adds two drops each of bergamot and geranium essential oil to the water and luxuriates in the bath as she feels her tense muscles unclench.

At bedtime, she takes the spray bottle into the bedroom and lightly mists her pillow and notices the lavender scent when she lays down. In a few minutes, she falls asleep and sleeps through the night. She rises in the morning refreshed for the first time in months.

Rhonda knows that much of her ability to control her anxiety symptoms comes from the medications her medical doctor prescribed, but she credits the aromatherapy with helping her lose her fear of relapse and undoing years of habits that kept her from encountering situations that triggered her symptoms. She continues to use the diffuser, the scented bath, and the lavender spray on days when she feels particularly anxious and sometimes just because she enjoys the scents themselves. Soon she finds herself feeling in control of her symptoms and begins venturing out to share social experiences with her friends.

Analysis

Rhonda succeeded in using aromatherapy in a constructive way to reduce the lingering symptoms of chronic anxiety. Her success can be credited to her understanding that her anxiety is a chemical disorder, not a weakness or a failing, and that aromatherapy can work in tandem with medication.

She took a route that involved expert advice from a practitioner who has earned a certification in aromatherapy from an accredited college and who keeps up with the most recent science through a national professional association. This gave Rhonda confidence that she would receive well-informed advice from a professional—one who did not sell the essential oils themselves. She knew that by seeing a therapist instead of a multilevel marketing sales rep, she had a much better chance of receiving sound advice that would actually help relieve her anxiety. Having the option of shopping for oils on her own also increased her confidence that she was buying the purest essential oils and that she had purchased only the oils she needed instead of an expensive "starter set" with products she might never use.

A paper published in *Evidence-Based Complementary Alternative Medicine* (Koulivand et al., 2013) reviewed all the studies to that date that examined the effects of lavender on neurological disorders. Many of these studies revealed that lavender has anxiolytic properties, meaning that the essential oil can have a direct, positive effect on anxiety disorders. This has been well known by practitioners of ayurvedic medicine for centuries, but now science backs up the observational evidence that inhaling lavender's scent or using it in aromatic massage can reduce anxiety in humans. Likewise, several recent studies have shown that clary sage (specifically *Salvia sclarea*) reduces blood pressure, respiratory rate, and stress in humans (Seol et al., 2013) and also acts as an antidepressant (Seol et al., 2010).

Geranium essential oil has demonstrated a powerful ability to reduce anxiety in women during labor, especially in those who have never delivered a baby before. A randomized, double-blind study published in the *Journal of Caring Science* (Fakari et al., 2015) found that inhaling the scent of geranium essential oil significantly reduces anxiety in women giving birth, arguably one of the most stressful events in a woman's life. According to several studies dating back to 2007, bergamot essential oil not only reduced stress in laboratory animals but also may have the ability to protect the nervous system from disease.

Finally, the scent of lavender is well known in the hotel industry for its ability to help people sleep. Some hotels actually distribute small spray bottles of lavender-scented water for guests to mist over their pillows or into the room, helping to promote rest and relaxation during their stay. A randomized, double-blind study published in *the Journal of Alternative Complementary Medicine* (Lillehei et al., 2015) found that wearing a lavender inhalation patch on the chest at night promoted better sleep quality, although it did not help the subjects get more hours of sleep than the control group. A literature review completed in 2020 and published in *Evidence-Based Complementary Alternative Medicine* (Guadagna et al., 2020) found that lavender was one of the most frequently studied and tested plant extracts and that studies that administered lavender in pill form (silexan) showed "significant improvement in sleep quality and anxiety compared to placebo."

Working with a certified aromatherapist gave Rhonda the advantage of using oils based on science rather than which oils will net a salesperson the greatest profit. She enjoyed true benefits from the use of aromatherapy as a complement to her medication, because of her therapist's expertise and credibility.

4. ALTERNATIVE OR COMPLEMENT?

Henry, a man in his late 20s, goes to the gym daily and often walks around the locker room in his bare feet. As a consequence of this, he picks up a case of *tinea pedis*, more commonly known as athlete's foot. This results in itchy, red patches on his feet and cracked, weeping skin between his toes.

Henry follows a natural lifestyle and has so far avoided the need for medications in his adult life. He calls his doctor to find out the best therapy for his infection, and the doctor recommends a topical antifungal cream like Lotrimin Ultra or Phytozine. Henry uses "Doctor Google" to study each of these remedies and decides that they are not natural enough for his tastes. He does further research and finds that thyme, cinnamon, and clove essential oils are said to have antimicrobial properties and are specifically recommended to fight athlete's foot and other forms of ringworm.

Using coconut oil as a carrier oil, Henry places a tablespoon of the coconut oil in a small bowl and mixes two drops of each of the three essential oils into it. He applies this generously to his feet and rubs it between his toes and then puts on a pair of clean white socks. He repeats this process several times a day.

After five days, Henry finds himself scratching his feet one evening, and he takes off his socks and looks at them closely. The redness, scaliness between his toes, and itching have not abated. He goes back online and looks for a specific recipe for a natural remedy for athlete's foot and finds that tea tree oil and bitter orange essential oil are also recommended for treatment of fungal infections. Again, he does not find clear instructions for how much to use or for how long, so he mixes two drops of each of these oils with coconut oil as a carrier and applies this to his feet three times a day. He finds that the itching subsides, but the scaly skin and redness do not.

Now Henry feels that he cannot call his doctor for advice, because he did not follow her advice the first time he called. He looks for additional ideas online and finally determines that he may need to give in and use an over-the-counter fungal cream. He buys a tube of the extra-strength cream and uses it twice daily as directed. In seven days, the athlete's foot clears up completely.

Analysis

The problem with home remedies is that there may be no clear guidance on dosage, as there is with pharmaceuticals and over-the-counter medications. All the essential oils that Henry used to attempt to clear up his fungal infection may actually have the power to do so, and some have demonstrated this ability in laboratory tests. There have been few (if any) tests on humans, however, so appropriate dosages for curing an infection with a human subject have not been established.

Henry used two drops of each essential oil and applied them several times a day with no success. Would four drops of each oil have been the definitive cure? Would six drops be required? He has no way to know, and various practitioners that publish their own recipes on their websites probably did not run randomized, double-blind experiments on large groups of people to find out if their recipes can actually cure an infection.

On top of the cloudy nature of the dosages in this case, Henry had no way to know how long it might take for the essential oils to work. Some antifungal pharmaceutical creams can take weeks to fully eradicate a fungal infection, so it might be that thyme and clove essential oils need weeks to work as well. How long should he continue the treatment if he sees no progress? This young man found himself in a quandary of his own making, with no plan to consult a professional in alternative therapies to get answers to his questions. Worse, he now felt foolish for trying to treat the infection with essential oils instead of taking his doctor's advice, so he did not feel that he could call to see if she knew anything more than Google did about antifungal essential oils. (His doctor probably would have been happy to discuss this with him, but Henry felt too embarrassed to make this call.)

It is conceivable that Henry could have used the essential oils in addition to the antifungal cream so that he could get the benefits of both. He would have a surefire way to kill his athlete's foot with a proven, well-established antifungal, as well as the odor-defeating, soothing powers of the essential oils when applied before the cream. Perhaps his experience will lead him to take this kind of action should he ever contract this nasty infection again—or maybe he would add a pair of shower sandals to his gym bag and avoid walking around barefoot in a public locker room, thus avoiding a recurrence.

5. LIFE OR DEATH

Daniel notices a suspicious mole on his chest just below his neck. He sees his doctor, who examines the mole and sends him to a dermatologist for a biopsy. Sure enough, the biopsy reveals that Daniel has melanoma, a skin cancer that can metastasize (spread) and can even be fatal. The dermatologist removes the mole and sends it to a lab to be sure the margins around the cancer are clean.

The lab results indicate that there may be more of the cancer in Daniel's chest, so he is referred to an oncologist, who begins traditional therapies including chemotherapy and radiation. Like many cancer patients, Daniel researches his disease thoroughly on the internet, often during the sleepless nights he spends worrying about the progression of his illness. He reads the protocols and sees that he is getting the best possible standard of care, but he continues to search for other things he can do to help ensure that the cancer does not get worse.

He comes across a study that suggests that frankincense essential oil killed breast cancer cells in vitro—that is, in a test tube in a laboratory. This result appears in another study as well, this time showing that frankincense essential oil inhibits the growth of bladder cancer cells. Neither of these studies involved tests on animals or humans, but Daniel quickly generalizes the results, deciding that if frankincense worked on two kinds of cancer cells, it would probably work on others.

He dilutes several drops of frankincense essential oil in a little jojoba oil and dabs this on the incision at the spot where the dermatologist removed the melanoma. Each day, he makes this part of his morning and evening regimen, adding the diluted frankincense oil to the healing wound and covering it with some gauze. He mentions this to his oncologist, who points out that this essential oil has not been tested on humans with cancer, so Daniel should not expect much result from this practice. However, the oncologist acknowledges that the frankincense oil won't do any harm, so Daniel continues to dress the incision with the diluted oil.

When the chemotherapy and radiation have been completed, testing concludes that the melanoma cells are no longer present. Daniel and his family rejoice at the news, and he proclaims his experiment with frankincense essential oil a success. "If it weren't for that study I found, I might still have cancer now," he tells his family and friends. "Frankincense essential oil really saved my life."

"What about all the chemo and radiation?" his wife asks. "You don't think those things had something to do with it?"

"Oh, maybe," he replies, "but the real cures are in what God gave us, in nature."

Analysis

Few people are more vulnerable to misinformation than those searching the internet in the middle of the night for lifesaving cures to a potentially fatal illness. Preliminary studies suddenly appear to reveal miracle cures—and in such scenarios, patients see themselves as helpless victims of the U.S. Food and Drug Administration (FDA), the regulating body standing between them and the panacea that could save their lives. This can lead some patients to take matters into their own hands, dosing themselves with unproven substances in the vain hope that they will provide wondrous results.

A study conducted in 2015 and published in the *Asian Pacific Journal of Tropical Biomedicine* produced a substance it calls "frankincense derived heavy oil" from plant resin, using the chemical hexane to extract the oil. The researchers then prepared different dilutions of the oil to change the amounts of various terpenes in it, to see which of these affected breast cancer cells in a test tube. This allowed the researchers to apply the oil directly to the cancer cells, to see if it inhibited their growth. What is important about this study is that it did not use frankincense essential oil against the cancer cells—the heavy oil has a different balance of terpenes. In fact, the researchers also tested frankincense essential oil on the breast cancer cells and concluded, "Very low concentration of . . . heavy terpenes elicits considerable cytotoxicity on MDA-MB-231 cells compared to hydro distillated essential oil derived from frankincense resin." In short, the heavy oil was much more effective than the essential oil in killing breast cancer cells in a test tube.

Daniel, of course, did not have breast cancer, nor could he apply the essential oil directly to the remaining melanoma cells in his body. The very preliminary study provided no instructions to people about how much frankincense might be required to kill cancer cells. Most notably, the study did not involve humans (or even mice, for that matter) and served only to suggest that this particular oil might have future therapeutic value. Much more research will be required before this phase I study can lead to the use of frankincense in cancer therapy.

None of these facts, however, could deter Daniel from his conviction that his use of frankincense saved his life—despite the highly reliable, well-documented effectiveness of the chemotherapy and radiation he had to endure for months. Fortunately, he did not refuse the medical therapies in favor of a natural approach, and this decision undoubtedly had more effect than the topical application of few drops of essential oil.

Daniel might have been better served to use essential oils to alleviate some of the unpleasant symptoms that traditional therapies can cause. Diffusion of essential oils can quiet feelings of nausea, improve sleep, promote relaxation, and deepen meditation and mindfulness, all of which can help make cancer treatment more bearable for patients.

Glossary

Adulterated: An oil that contains alcohol, synthetic chemicals, water, or anything but pure, 100 percent essential oil.

Analgesic: A non-narcotic substance that relieves pain. Aspirin, ibuprofen, and acetaminophen are common analgesics.

Anxiolytic: Capable of relieving anxiety and stress.

Aphrodisiac: A food, scent, or other substance that promotes sexual desire.

Aromatherapy: The use of scented materials such as essential oils, incense, or plant matter for medicinal benefits.

Astringent: A substance that makes skin or other body tissues tighten and contract, reducing discharges like oils. Astringents are often used to combat acne.

Carminative: A natural substance or a medicine that prevents gas from forming in the intestinal tract.

Carrier oils: Oils used to dilute essential oils. Most essential oils cause irritation if they are applied directly on skin, so they need to be diluted in a carrier oil before application. In addition, carrier oils hold their liquid form when heated, so they do not evaporate the way essential oils do.

Cephalic: A substance that clears fogginess from the head, enabling the user to think more clearly.

Chain of supply: The line of growers, harvesters, distillers, bottlers, labelers, marketers, therapists, and salespeople between the plants that provide essential oils and the end user.

Constituents: The natural chemicals that give the essential oil its therapeutic value. Every oil can be broken down into individual chemicals, some of which have more ability to fight microbes or perform other therapeutic tasks than others.

Diffuser: A device that heats the essential oil and allows it to evaporate into the air, where users can inhale it and enjoy its benefits.

Dilutants: Synthetic additives some bottlers add to their essential oils to lower the cost to the consumer and to extend the supply of a specific oil.

Distillation: The use of steam and/or water to extract the oil from a plant by breaking down its petals, leaves, stems, flowers, roots, or bark.

Diuretic: Any substance that increases the production of urine.

Double-blind: A study in which neither the researchers nor the subjects know which subjects received the substance being tested and which received a placebo. This prevents bias in the results.

Drop: While the size of a drop of essential or carrier oil varies based on the dropper, it is generally considered to be 1/600th of a fluid ounce.

Essential oil: A liquid extracted from plant material containing the plant's essence, defined as its scent.

Expectorant: A medication or other substance that encourages nasal or chest congestion to clear.

Expression: A process in which a fruit's rind is rotated and punctured to release and collect the juice and essential oil.

Febrifuge: A medication or substance that lowers body temperature, specifically to alleviate fever.

Fixed oil: A natural oil that does not change its state when heated.

Humidifier: A mechanical or electronic device that generates a mist, to add moisture to the air in a room.

In vitro: Laboratory research conducted on cells or tissues in test tubes or Petri dishes, with no involvement of animal or human subjects.

In vivo: Research conducted on animals or people.

Nature identical oil: A misleading label that indicates that the oil is synthetic, created in a laboratory with chemicals, dilutants, and other adulteration.

Photosensitization: Sensitivity to sunlight, with the potential to burn or blister human skin. Citrus oils are photosensitive and should not be used on skin before going out into the sun.

Pipette: A glass tube used to measure tiny amounts of essential oil.

Placebo: A substance (a pill, liquid medication, injection, or otherwise) that contains no medicinal compounds, used to simulate treatment in double-blind experiments.

Pure: An essential oil is pure when it contains nothing but 100 percent essential oil.

Solvent extraction: Use of a solvent such as methanol, ethanol, hexane, or petroleum ether to extract scented molecules from a plant or resin. Frankincense, for example, is extracted from a resin using hexane.

Terpene: The organic compound a plant produces that contains its scent. Essential oils all contain terpenes.

Therapeutic grade: A term made up by the essential oil industry to imply that essential oils are graded and regulated and that some manufacturers' products are better than others. The essential oil industry is not regulated by any federal agency, so there are no standards used to produce a "therapeutic grade" of any essential oil.

Topical: Used on skin, on the outside of the body.

Vaporizer: A mechanical device that creates steam.

Volatile organic compounds (VOCs): VOCs are chemicals with a low boiling point and the ability to release vapors at room temperature, most of which are manufactured. They are emitted as gases from liquids or solids, so most chemical products with any kind of scent contain VOCs, most of which can be harmful to people. Essential oils are also VOCs, though most of these do not have the potentially harmful effects of solvents, transportation fuels, and many others.

Directory of Resources

BOOKS

Axe, Josh, Rubin, Jordan, Bollinger, Ty. *Essential Oils: Ancient Medicine.* Destiny Image, Shippensburg, PA, February 2018.
Written in part by a medical doctor, this book provides credible advice about the use of essential oils as natural alternatives to prescription medications, household cleaners, and personal care products.

Cohen, Jodi. *Essential Oils to Boost the Brain and Heal the Body: 5 Steps to Calm Anxiety, Sleep Better, and Reduce Inflammation to Regain Control of Your Health.* Ten Speed Press, Berkeley, CA, March 2021.
In a well-organized and easy-to-use volume, Cohen offers evidence-based approaches to a number of common health issues including chronic inflammation, stress and anxiety, and mood disorders.

Minetor, Randi (uncredited). *Essential Oils and Aromatherapy: An Introductory Guide.* Sonoma Press, Emeryville, CA, 2014.
My own book provides a basic primer to a wide range of essential oils and their origins, proper methods of mixing and applying them, and plenty of recipes to help beginners master the use of essential oils safely and effectively.

Minetor, Randi. *Essential Oils of the Bible: Connecting God's Word to Natural Healing.* Althea Press, Emeryville, CA, July 2016.

Extending the use of essential oils to spiritual practice, this book links today's oils to the original plant matter and oils expressed from these plants in biblical times, with direct references to specific uses for the oils mentioned in the scripture. It covers both the Old and New Testaments to help readers extend their own relationship with God by using scents and aromatherapy in worship.

Tiwari, Maya. *Ayurveda: A Life of Balance: The Complete Guide to Ayurvedic Nutrition and Body Types with Recipes.* Healing Arts Press, Rochester, VT, December 1994.
A practical guide to balanced living through the ancient principles of Ayurveda, this book offers a cogent and thorough profile of the use of foods, home remedies, and other elements to achieve physical, spiritual, and emotional health. While not specifically about essential oils, it provides an understanding of the benefits of Ayurvedic practice in which oils do play a significant role.

Worwood, Valerie Ann. *The Complete Book of Essential Oils and Aromatherapy.* New World Library, Novato, CA, November 2016.
Updated and revised in 2016, this seminal handbook for essential oils users and practitioners provides more than 800 recipes for health and beauty, including many blends to be used in aromatherapy. Worwood holds a doctorate in complementary health, and she teaches clinical aromatherapy and practices it with her own clients.

ORGANIZATIONS

Alliance of International Aromatherapists, Thornton, CO, www.alliance-aromatherapists.org
AIA is dedicated to the education of aromatherapists, health care professionals, and the public in all aspects of aromatherapy. It serves as a network of practitioners who "collectively strive to advance the profession of aromatherapy and to serve the public."

American Botanical Council, Austin, TX, www.herbalgram.org
This nonprofit organization educates consumers, health care professionals, researchers, educators, and the media on the responsible use of herbs and medicinal plants, their basis in science, and their place in the world of health. It also publishes *HerbalGram*, a peer-reviewed quarterly science journal to increase public awareness about medicinal herbs.

National Association for Holistic Aromatherapy, Pocatello, ID, naha.org
This nonprofit organization works toward the integration of aromatherapy into complementary health care practices, including home

pharmacy and self-care. It offers the general public the most current scientific information about essential oils and reaches aromatherapy and holistic practitioners, businesses, writers, educators, health care professionals, the media, and designers of aromatherapy products.

WEBSITES

AromaWeb, operated by AromaWeb LLC, www.aromaweb.com
Separating itself from the information published online by essential oil marketing companies, AromaWeb provides objective, "brand-neutral" information; guides; lists; articles; blends; recipes; book reviews; and more. Its website serves as a clearinghouse for all manner of information about essential oils and aromatherapy.

"Aromatherapy." U.S. Food and Drug Administration, www.fda.gov /cosmetics/cosmetic-products/aromatherapy
This comprehensive website answers many common questions about aromatherapy, essential oils, their safe use, the advertising claims made about these oils, and regulatory issues. It provides the most up-to-date statements by the FDA about the use of essential oils in consumer markets.

"Part 182: Substances Generally Regarded as Safe." U.S. Food and Drug Administration Code of Federal Regulations Title 21, Chapter 1. www.accessdata.fda.gov/scripts/cdrh/cfdocs/cfcfr/CFRSearch .cfm?fr=182.20&SearchTerm=peppermint
The extensive GRAS list includes a specific section about plant substances and which are safe for their "intended use," which may or may not include consumption. This provides a starting point for research about whether or not you may use an essential oil in food or in contact with the human body—if the oil is not on this list, it is best to avoid using it.

PubMed.gov
Search on "essential oils" at this open-access database to find abstracts (and often the full text) of scientific studies on single, blended, and comparative varieties of essential oils. The information is available here for free from life science journals at the National Institutes of Health's National Library of Medicine, where most studies around the world are catalogued. While this vast database contains studies on just about every aspect of biomedical research, it does contain thousands of studies on specific essential oils and broader research that compares the performance of a selected oil against others to test effectiveness in a controlled situation.

Science Direct—Essential Oils: www.sciencedirect.com/topics/agricultural
-and-biological-sciences/essential-oils
This website provides a rolling catalog of continuously updated cita-
tions and abstracts of recent peer-reviewed, published studies in journals around the world.

Index

About the Author

Randi Minetor, MA, is a medical journalist and the author of *Essential Oils and Aromatherapy: An Introductory Guide* and the bestselling *Essential Oils in the Bible*. For Greenwood, she wrote *Debating Your Plate: The Most Controversial Foods and Ingredients*, *Medical Tests in Context: Innovations and Insights*, and *Blowing Up: The Psychology of Conflict*. Her extensive book list as an author includes more than 60 books on nature, travel, and general interest topics. She writes for *Western New York Physician* magazine, and she has served as a principal writer of patient, consumer, and doctor-to-doctor materials for the University of Rochester Medical Center. She is a graduate of the University of Rochester and the University at Buffalo. Follow her on Facebook @minetorbooks, on Twitter @rminetor, and on Instagram @writerrandi.